The Economics of Information, Communication, and Entertainment

The Impacts of Digital Technology in the 21st Century

Series Editor

Darcy Gerbarg, New York, NY, USA

For further volumes:
http://www.springer.com/series/8276

Judith O'Neill • Eli M. Noam • Darcy Gerbarg
Editors

Broadband as a Video Platform

Strategies for Africa

Springer

Editors
Judith O'Neill
CMAS Holdings LLC
New York, NY
USA

Eli M. Noam
Columbia Business School
Columbia University
New York, NY
USA

Darcy Gerbarg
DVI, Ltd and
Senior Fellow, CITI
New York, NY
USA

ISSN 1868-0453 ISSN 1868-0461 (electronic)
ISBN 978-3-319-03616-8 ISBN 978-3-319-03617-5 (eBook)
DOI 10.1007/978-3-319-03617-5
Springer Cham Heidelberg New York Dordrecht London

Library of Congress Control Number: 2013957932

Printed on acid-free paper

Springer is part of Springer Science+Business Media (www.springer.com)

Foreword

At the ITU, we believe that information and communication technologies are a key enabling platform for innovation and development. This collection of thoughtful contributions by leading experts, edited by Judith O'Neill, Eli Noam, and Darcy Gerbarg, explores the historical, economic, social, and regulatory context why African broadband penetration is low and offers their perspectives on how to bring more Africans to the global information society.

What is the actual situation of broadband in Africa? Most of us know that the African region[1] has experienced strong growth in mobile cellular network and service developments since 2008, allowing an increasing number of the almost 850 million people in the region to get connected and join the information society. While the African region has extremely limited fixed (wired) ICT infrastructure, the progress that has been made in terms of mobile cellular uptake and growth is nothing short of remarkable. Within five years, the region's mobile cellular penetration rate has doubled from 32 to 64 % and active mobile broadband penetration, which was for all purposes inexistent in 2008, has grown to almost 11 %. At the same time, the number of Internet users has grown steadily, from 6 % in 2008 to 16 % in 2013. Close to 7 % of households in Africa now have Internet access at home, compared to only 2 % in 2008.

Yet despite this progress and the impact that mobile cellular services have had in Africa, the region's ICT penetration levels remain below global and developing-country averages. The region lags particularly in expanding access to the fixed telephone network and to high-speed fixed broadband services. Fixed telephone penetration has remained low (at below 1.5 %), and the number of fixed (wired) broadband subscriptions per 100 inhabitants stands at 0.3 in 2013, well below the global average of 9.8 % (see below).

[1] The list of countries included in the African region is based on the country grouping of ITU BDT regions; see http://www.itu.int/ITU-D/ict/definitions/regions/index.html.

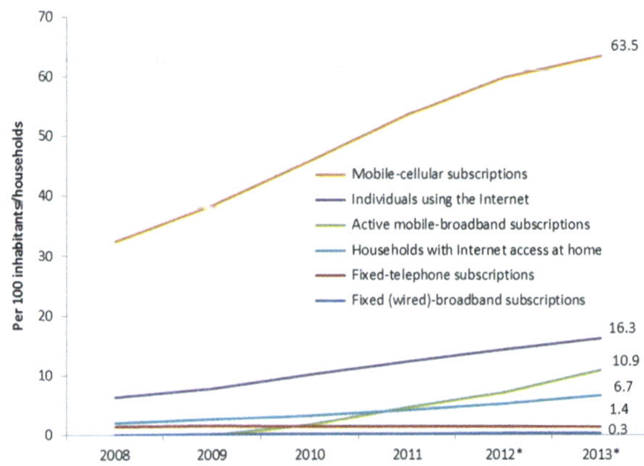

Source: ITU World Telecommunication/ICT indicators database.
Note: *Estimates

Both the percentage of the population using the Internet and the proportion of households with Internet access have been growing since 2008, at 27.4 and 20.8 % compound annual growth rates, respectively. A recent slowdown in growth in Internet user penetration levels (which has reached 16.3 % at the end of 2013) despite strong growth in the number of active mobile broadband subscriptions (see below) suggests that access and infrastructure may not be the only barrier to more Africans coming online.

Although fixed broadband penetration is growing faster in Africa than in other regions, penetration levels remain low. Overall growth rates in fixed broadband penetration have decreased, from almost 40 % in 2009/2010 to below 30 % in 2012/2013, suggesting that the current average penetration rate (below 1 % in the region) is not going to substantially change soon. A limited fixed telephone network that has been showing negative to no growth (which is also the case in many developed countries) is part of the challenge in bringing fixed (wired) broadband services to a larger population and more regions (see below).

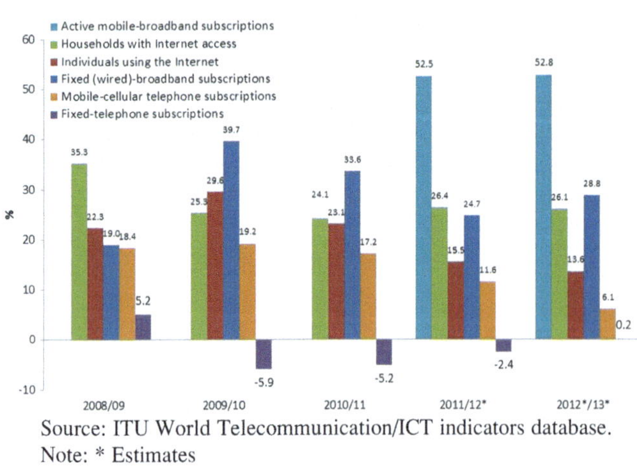

Source: ITU World Telecommunication/ICT indicators database.
Note: * Estimates

The number of fixed (wired) broadband subscriptions in Africa remains extremely low, at an estimated 3 million by the end of 2013. The region's fixed broadband penetration rate of 0.3 % lies below the world (9.8 %) and developing-country (6.1 %) average. It is also low compared to penetration rates in other developing regions of Arab states (3.3 %), Asia and the Pacific (7.6 %), the Commonwealth of Independent States (CIS, 13.5 %), and the Americas (17.1 %).

Within Africa, a handful of countries have fixed broadband penetration rates of above 2 % and only the island states of Seychelles and Mauritius have penetration rates that exceed the global average. Key barriers to higher fixed broadband penetration are the limited availability of fixed telephone networks and the high cost for fixed broadband services. In a number of developing countries, an insufficient backbone infrastructure and lack of access to international Internet bandwidth are also limiting the delivery of high-speed Internet access (see below).

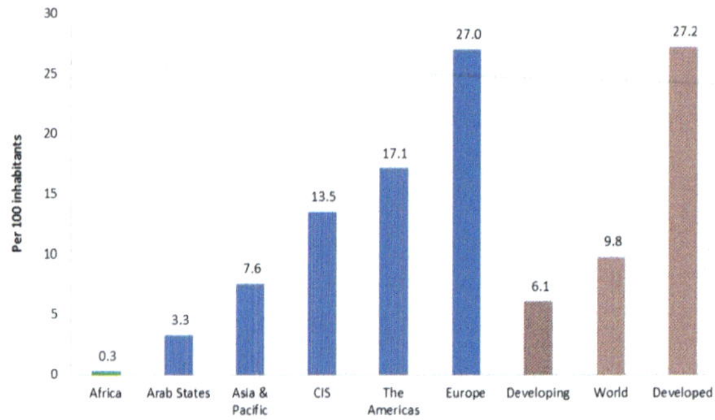

Fixed (*wired*) broadband subscriptions in Africa, by region and level of development, 2013* (*top*), and by country, 2012 (*bottom*). Source: ITU World Telecommunication/ICT indicators database. Note: *Estimates. The *red line* in the *bottom* chart shows the 2012 world average

Will mobile broadband be the *broadband savior* for Africa? ITU estimates that the African region will have a total of 93 million mobile broadband subscriptions by the end of 2013, some 33 million more than the year before. At 10.9 %, regional mobile broadband penetration is below the global (29.5 %) and developing-country (19.8 %) average with many opportunities for growth (see below).

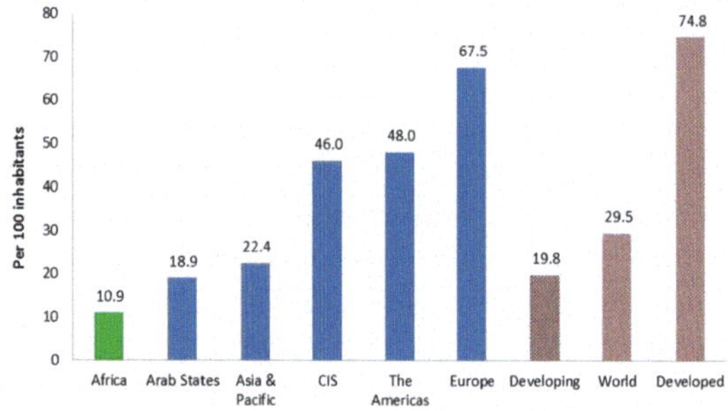

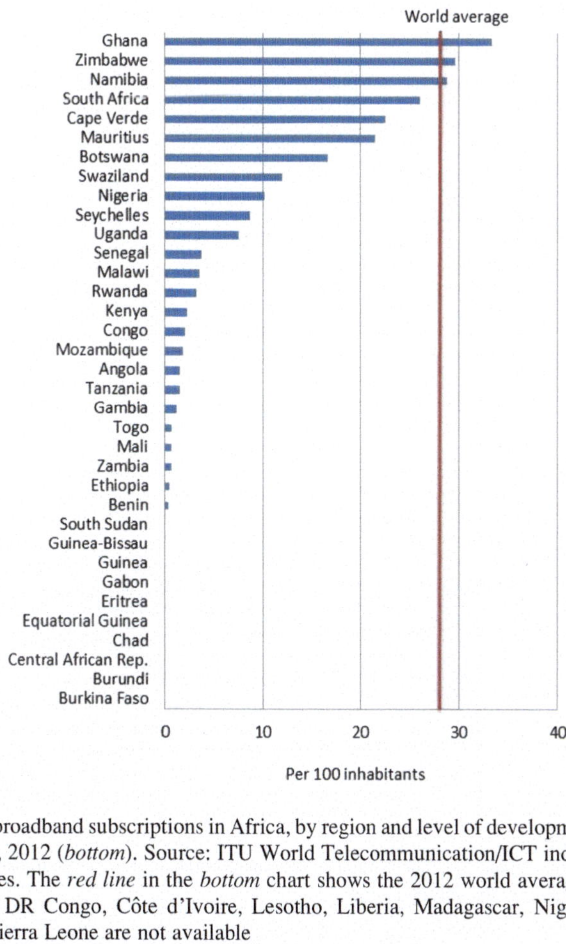

Active mobile broadband subscriptions in Africa, by region and level of development, 2013* (*top*), and by country, 2012 (*bottom*). Source: ITU World Telecommunication/ICT indicators database. Note: *Estimates. The *red line* in the *bottom* chart shows the 2012 world average. In the *bottom chart*, data for DR Congo, Côte d'Ivoire, Lesotho, Liberia, Madagascar, Niger, S. Tome and Principe, and Sierra Leone are not available

Ghana, Zimbabwe, Namibia, and South Africa have the region's highest 2012 mobile broadband penetration rates, of above 25 %. The island states of Cape Verde and Mauritius have also made good progress and reached 21.5 and 22.5 % of mobile broadband penetration. At the same time, a large number of African countries were late to launch mobile broadband networks or have yet to launch 3G high-speed networks. Thus, wireless broadband penetration is low in many countries, and more than half of the African countries had a penetration of less than 5 % by the end of 2012.[2]

The indicators above indicate that there are many opportunities for improvement and we need to redouble our efforts to bring broadband to all of Africa's citizens. The time is ripe for novel ideas for how to achieve this combined with "out-of-the-box" innovative approaches. The following collection of essays by leading experts is an important contribution to this complex debate.

Geneva, Switzerland Brahima Sanou

[2] For some countries, including DR Congo, Lesotho, Liberia, and Niger, data on active mobile broadband subscriptions are not available.

Preface

This book contains information, data, research, observations, personal interviews, and decades of telecommunications and new media experience of experts working in Africa. It memorializes and brings the current contributions by African university professors and telecommunications enterprises, leading telecommunications and media private sector consultants, and seasoned experts in multilateral institutions such as the World Bank and the International Telecommunications Union as well as AT&T and Ericsson, two of the world's major carriers.

The conference that this volume embodies was titled "Next Generation Broadband as a Video Platform—Strategies for Africa" and took place in Lusaka, Zambia, in May of 2012 over a period of two days. The chapters in this volume reflect the topical exchanges that occurred among the experts; members of the Zambian Government, who graciously hosted the Conference; members of the Zambian telecommunications and media sectors; and followers on Webex around the world. The information from the conference has been updated and edited for inclusion in this book, and a few additional articles have been added.

Most of the experts who participated in the Conference agree that, if made reasonably available in most African countries, next-generation broadband, with or without a video platform, would have a shallow learning curve and a fast uptake. However, it is logical to assume that primarily due to the lack of landline infrastructure throughout the continent the next-generation broadband growth in Africa will be based on mobile penetration. Thus, there is a note of sobriety expressed by some of the authors that, while feasible, and even happening in certain countries, it seems likely that next-gen broadband in Africa will be mobile and thereby take longer and require more technology and pricing ingenuity than the more traditional growth patterns experienced by industrialized nations. After all, most industrialized nations had the benefit of starting their next-gen build-out on a landline backbone. The statistics presented in the chapters by Tim Kelly, Drs. Gillwald and de Lanerolle, Professor Hudson, Mr. Akoh, and others in this volume bear out these observations.

Perhaps the most interesting and notable result of the conference, reflecting the millions of work hours dedicated to the topic in their careers and presented papers, and the resulting chapters herein, is that there are still vast discrepancies in Africa's prognosis for keeping up with, or catching up to, a world of competitive, ever-increasing broadband speeds and access. What that means to a nation's development, where even the definition of broadband itself is not fully agreed upon by the nations in the competition, is hotly debated. The chapters in this book provide different vantage points, perspectives, and experiences leading to varying conclusions of Africa's likely fit in the next-generation broadband world.

The Introduction which follows this Preface puts the relevance of the history of Africa's telecommunications sector in perspective, as a key factor in Africa's next-generation broadband. Judith O'Neill has spent nearly 40 years of her career in telecommunications and new media working, advising, and seeking creative, workable strategies in developing countries, much of the last 25 of which were in and around most of the 54 countries of sub-Saharan Africa. The perspective which she presents is at once legal and regulatory, cultural and pragmatic, and to a lesser degree political.

We hope that this volume fills in some of the holes in your research on African broadband, particularly the prospects for next-generation broadband in Africa and its role as a video platform. It should provide an insight into possibilities and challenges for Africa in this field. We hope that it allows you to put your own background, thoughts, and work into context so that you can better understand Africa's road to broadband ubiquity as a continent and next-generation broadband as a strategy for catching up with the world economy.

Finally, we wish to acknowledge and express our sincere gratitude to those individuals and institutions without whom and which the Conference in Zambia and this resulting volume could not have happened. We thank our core private sector sponsors, AT&T and Liquid Telecoms, for their moral, fiscal, and technological support before, during, and following the conference as well as Airtel and Emerging Markets Communications, EMC, for their substantive and generous contribution to our program. We also thank our gracious and committed hosts, the Government of Zambia and in particular the Ministry of Transport, Works, Supply and Communications and especially the Honorable Minister Yamfwa Mukanga. We thank as well ZICTA and its tireless Director General Margaret Chalwe Mudenda for her enthusiasm, the protocol assistance of ZICTA with visiting dignitaries, and the richness of her vision and presentation on Zambia's telecommunications future. We thank our Zambian coordinators, Shuller Habeenzu and Charles Chilinda, who provided us the insights to ensure that our program ran smoothly in Zambia from start to finish.

Of course neither the conference nor this volume could have existed without the time, energy, and life commitment to their career expertise in all aspects of telecommunications services and next-generation broadband of our speakers, academics, private sector leaders, and multilateral experts.

Last, but certainly not least, we express our gratitude to CITI for its vision and to Jason Buckweitz for a sleepless 52 h in Lusaka to ensure that our program ran technologically smoothly in all aspects; and we applaud and thank Mr. Brahima Sanou, Director of the Development Bureau of the International

Telecommunications Union, for being our keynote speaker and author of the Foreword to this volume, for his sincere and lifelong commitment to the improvement of Africa and its citizens through telecommunications, and for extending the torch of next-generation broadband as a guiding light to Africa.

New York, NY, USA

Judith O'Neill
Eli M. Noam
Darcy Gerbarg

Contents

Chapter 1
Introduction

Judith O'Neill, Esq

The History of Africa as It Moves from State-Owned to Private Telecommunications Enterprises and Its Impact on the Deployment of Next-Generation Broadband as a Video-Capable Platform in Africa.

While there are large pockets of fiscal comfort in sub-Saharan Africa, the vast majority of the continent's consumer-aged population is cash strapped, investment cautious, and lives in countries presenting real and imagined challenges to foreign direct investment. It is also not a secret that next-generation, or any generation, broadband in Africa requires considerable up-front capital investment to build infrastructure that currently does not exist. Thus, without an existential need, extraordinary potential for market uptake, and content generation, the thought of deploying next-generation broadband in most of sub-Saharan Africa would be a nonstarter. The view of this author, having spent more than 20 years in virtually every aspect of sub-Saharan African telecommunications and media sectors, is that the need is indeed existential and the potential is indeed extraordinary. The best path to get there, however, is not entirely clear.

Thus, the direct question is as simple to ask as it is difficult to answer. What is the best approach for the governments of the 54 countries of sub-Saharan Africa to take to encourage the private sector to "make it happen"? The presumption here is that for success, two things will happen first. The government will be in a facilitating, regulatory role, monitoring the behavior of stakeholders in the sector and the private sector will make sober evaluations of the investment environment. These evaluations will be based on demographics, economics, transparency of the rules

J. O'Neill, Esq (✉)
Chief Operating Officer, CMAS Holdings LLC, Two Columbus Ave. 26th Floor, New York, NY 10023, USA
e-mail: joneill@cmasalert.com

J. O'Neill et al. (eds.), *Broadband as a Video Platform*, The Economics of Information, Communication, and Entertainment, DOI 10.1007/978-3-319-03617-5_1,

and process, government's means of facilitation, and long-term prospects for success. Evaluating the appropriate breadth and scope of investment capital needed while never easy is sure to be particularly confounding and complex in Africa.

A useful comparison for considering the state of telecommunications infrastructure in sub-Saharan Africa is to look at Latin America, which began its process of seeding investment in telecommunications infrastructure in the 1980s. While Chile was the trailblazer in getting private investments in telecommunications infrastructure on the continent in the 1980s, many South American countries undertook the privatization of their national telephone company in the first half of the 1990s. Argentina, Peru, and Venezuela did transactions that raised billions of dollars in the early 1990s, accompanied by Mexico in North America which, in the last days of 1990, did a nearly $2 billion investment deal in its telecommunications sector. These transactions were the sale of majority stock ownership or operating control, or both, in the national government-owned telephone company. Sometimes deals included a guaranteed monopoly for a period of years, which in turn was coupled with a contractual, license-based operator obligation to invest huge amounts of money in national infrastructure and comply with minimum connectivity requirements for rural and poor urban areas. While this model took time for competition to develop particularly in landline services it did force at least some level of national infrastructure build-out in the country. It guaranteed a baseline of infrastructure in landline networks that would be important for later growth in the broadband sector.

Sub-Saharan Africa came later to the path of private investment in telecommunications and was more cautious about the introduction of private investment. With the exception of South Africa, African governments typically began allowing private investment in the late 1990s and the first half of 2000 and then only in discrete areas in telecommunications, predominantly the mobile sector where the landline monopolies had either not deployed or had deployed with little enthusiasm. It was not until 1997 that the Government of South Africa sold a 30 % stake in the national telephone company, Telkom, which gave the purchaser, SBC of the US and Malaysia Telecom, operating control of Telkom. Telkom was the nation's and the continent's largest telephone company. The payment of more than a billion US dollars for a minority stake in this company instigated the build-out in landline infrastructure that made Telkom one of the leading Internet, and eventually broadband, operators of Africa. While the investors purchased only 30 % of Telkom, they were given operating control, in a fashion similar to Latin American telecommunications privatizations.

The structure of the Telkom deal allowed the purchasers to invest a considerable amount of the high purchase price in the network and build-out. It carried with it as well obligations to commit to a certain level of infrastructure building. As Telkom was a key shareholder of, but did not operate the then main mobile company, Vodacom, the investment went into Telkom's landline assets. With that backbone, Telkom eventually became the nation's largest Internet company. Indeed, Telkom was poised, before a series of unfortunate decisions in 2005–2007, to deploy IPTV and had control of the landline network, a 50 % stake in the largest mobile operator, and owned the largest Internet business in the country. Though considerable advantages exist in the other nations of the continent that do not have the level and breadth of network infrastructure that exists in South Africa, even in South Africa, next-generation broadband is an investment challenge today, especially since the Telkom privatization. This is remarked upon with surprise by the editors of mybroadband.co.za.

Mybroadband.com is a broadband, tech, and IT news website whose focus is on the development of different speeds of connectivity in different countries and different regions of the world. On 25th January 2013, it published an article entitled "Surprising average mobile internet speed in SA" in which it included a chart of Internet connection speeds taken from "Akamai: The State of the Internet" report which shows South Africa with an average mobile Internet connection speed of 2.8 Mbps which was the lowest speed on the world chart. The commentator noted that this is surprising because South Africa is considered to be one of the nations in the forefront of mobile technology.[1]

Elsewhere in Africa emphasis was on mobile licenses, some niche Internet or other services considered value added at the time, and some attempts at investment in the government-owned telephone company. As the mobile penetration quickly began to outstrip the landline penetration, in what became known as the "mobile miracle" in Africa, privatization of national telephone companies became less relevant, and even Nigeria, which attempted three times to sell control of Nitel without real privatization success, ended up with a telecommunications sector that while competitive and robust consists mainly of mobile services with niche competitive landline operators and relatively little in the way of rural services. Unable or unwilling to sell their national telephone companies and thus garner the promise of future profitability that Latin America had, most African governments had little ability to demand the build-out obligations that Latin America was able to achieve through investor licensing agreements.

This is perhaps why the picture of infrastructure deployment in Latin America today is quite different from that in Africa. Where a private investor makes hundreds of millions or billions of dollars of investment in a national telephone company and takes control of it, its incentive, and its licensing mandate, is to build out infrastructure throughout the country, including downtrodden urban and rural areas. The Latin-style privatizations often had formulas of how much connectivity was required in many rural areas by a certain date to avoid breaching the license. In Africa, where private investment was not sought, the typical national operator had no such robust national requirements, and with very few exceptions, none were or could be levied. As a result the private investors built the networks that served only the particular service their license permitted them to provide. Thus, Africa today

[1] Mybroadband report, http://mybroadband.co.za/news/cellular/69212-surprising-average-mobile-internet-speed-in-SA, including Akamai: The State of the Internet chart of countries and regions around the world with respective mobile Internet speeds. There are likely multiple reasons for the speed of SA Telkom's mobile broadband. One may be the high cost of broadband in Africa relative to the rest of the world and the disproportionately high percentage of GNI it represents, as noted in Professor Nana Nzepa's chapter and noted below in a bar graph of relative costs of mobile broadband in Africa in this editor's chapter.

has a patchwork infrastructure built by individual licensees of individual services for the purpose of the discrete service they provide. The national telephone company often lacked the ability to invest in new infrastructure and thus keep up with technology. Even in South Africa, mobile penetration today is close to 100 % while landline penetration remains in single digits at a bit over 9 %. Throughout sub-Saharan Africa, general landline penetration ranges from 1 % or less to South Africa's percentage. This has a considerable impact on the strategy for deployment of broadband or next-gen broadband and video services.

This is shown in the statistics collected by the ITU and incorporated into the Foreword to this volume by Brahima Sanou, Director of the Development Bureau of the ITU. Internet penetration in sub-Saharan Africa as of 2012, which averaged about 16 %, was less than half of all other regions of the world including North Africa.[2]

As of 2013 sub-Saharan Africa still has a considerable gap in Internet penetration before broadband is introduced. This is likely due to the landline verses mobile infrastructure issue that negatively affects the region where landline is the starting point for the Internet and is largely nonexistent in Africa. Thus, when statistics on broadband are examined in the ITU reports only 7 of the 54 countries of sub-Saharan Africa show up on the comparative chart of fixed broadband subscriptions worldwide measured at 2 Mbps or more. Only Uganda, Senegal, Namibia, Cote d'Ivoire, and Mauritius showed up on the chart with speeds considered by many to be real broadband, with Kenya and Benin showing broadband with under 2 Mbps.[3] This likely has to do with the lack of fixed infrastructure in Africa as noted above but also with the added complication that the African model of purchasing mobile service, where the broadband growth currently exists, uses prepaid service rather than postpaid subscriptions. Approximately 80 % of the mobile service is prepaid.[4] The statistics would be different if mobile broadband were measured in sub-Saharan Africa, as we know from the inclusion of South Africa at 2.8 Mbps in the Akamai State of the Mobile Internet report cited above on mobile Internet speed.

[2] Brahima Sanou, Director of the ITU Telecommunications Development Bureau, *In the World 2013, ICT Facts and Figures, ITU publications*. Mr. Sanou was a keynote speaker at the Zambia conference on Next Generation Broadband as a Video Platform, which this volume memorializes.

[3] Id.

[4] Many factors in sub-Saharan Africa contribute to the prepaid model. One is that in some countries, fixed addresses are not prevalent, making postpaid services not feasible for invoicing. Also, family budgets are better accommodated by the up-front purchase of a fixed-fee mobile card, allowing a certain amount of minutes or capacity, rather than receiving an invoice at the end of the month that may not be anywhere within the budget. More recently, prepaid mobile cards have been adapted to and allow "mobile banking" which allows payments by African citizens who do not have a bank account (or a permanent fixed address). The prepaid phenomenon is not unique to Africa, however, as the vast majority of Latin American and Caribbean mobile service is prepaid, while broadband penetration outstrips sub-Saharan Africa, leading to a preliminary conclusion that the lack of fixed infrastructure in Africa and the increased expense and difficulty of providing broadband over mobile handsets are the key culprits to Africa's statistics in broadband.

Another factor in Africa is the cost of broadband there as compared to the rest of the world. As Professor Nana Nzepa points out in his chapter, the higher prices that Africa pays for communications services further expand its gap with industrialized countries. ITU statistics bear out the claim of disparity in the cost of mobile broadband service to a sub-Saharan African country versus prices in the industrialized world or even versus the world average. Indeed, the ITU report goes on to point out that these prices represent 36–58 % of GNI in the developing world to obtain a broadband connection, versus 1–10 % in the rest of the world.[5]

Broadband build-out in the industrialized world grew and found its market, applications, and business models using landline technology where the cost and infrastructure make access more ubiquitously available. While broadband is moving to mobile all over the world, the advantages of landline Internet that was elevated to the capacity of broadband, and now next-generation broadband, were born on earth and not in the airwaves. This creates some of the Internet penetration disparity between Western continents and many of the countries of Africa to which Drs. Gillwald and de Lanerolle refer in their chapter in this volume.

Taking next-generation broadband and applying it as an audio, and eventually a video platform in the air, as would be the logical progression in Africa, is a more challenging and costly exercise, though not insurmountable as Professor Hudson's study in interactive ICT/Internet farming points out. The introduction of spectrum allocation and other technological challenges adds additional complexity to determinations such as return on investment, no doubt making these determinations more difficult than they already are for landline Internet service. In the latter, the investment model we are seeing today is difficult enough: predicting consumer appetites, where and how consumers wish to view content, where content will come from, and who the next guerilla competitor will be. In Africa, accommodating a next-gen broadband service to mobile phones is added to these challenges.

Fortunately, as noted in the Hudson study, there are eagerness and market need for the services and applications that broadband, as an interactive audio and video platform, can bring. If deployed using the ubiquitous mobile platform in Africa, the combination of familiarity with the platform, extreme need, and excitement about the contribution of the interactive information being provided, as well as pride in entrepreneurship that Professor Hudson reports, is a great advantage in this market priming it for expansion.

This indicates once again that trying to find the "killer app" or business model that will command the investment required to advance basic Internet services to broadband and next-generation broadband in Africa may be putting the proverbial cart before the horse. As the Hudson study demonstrates, based on radio broadcasting alone, before a video platform is introduced, the magic is in the ability to interactively communicate. This is only enhanced by the addition of video and any number of additional apps. Focusing on the African need, desire and willingness to invest in interactive communication and using that as the springboard to

[5] Brahima Sanou, Director ITU Development Bureau, *The World in 2013: ICT Facts and Figures.*

development of interactive audio and video products for Africa may be a workable business approach for next-gen broadband in Africa. While as Professor Noam correctly indicates, the road is easier, less costly, and less technologically complicated if next-gen broadband is done on landline rather than mobile networks, this may not be an option for Africa, or at least the financial model of building the infrastructure will be a challenge. Thus, remaining agile to launch on mobile networks, while building out where feasible, landline facilities, may be the best course for sub-Saharan Africa.

The "Killer Policy" to Ensure the Success of the "Killer App" Whatever It May Be

An American educator named Richard Close, who examines community behavior in Africa, makes the point that it is not entertainment or any other individual "app" that will drive market uptake and thus investment in Africa, but rather the human need to communicate, one person with another on all topics of life from the high commercial to the most social.[6] The need of Africans, as the rest of us, to talk to and text and blog with each other may be the driver of investment in next-gen broadband in Africa, rather than any subsidiary application. Thus simple governmental policies that support interpersonal communications on a national basis, possibly in shared locations as required, and possibly starting with connectivity or shared infrastructure regulations, are candidates for first steps to a successful next-gen broadband program. This makes some sense and may be the answer hiding in plain sight in Africa, especially for those of us who take the existence of Internet landline backbones for granted.

In a nation where a small fraction of a percent of the population had the ability to communicate electronically with each other, Zimbabwe launched its first mobile service in 1998. It is a historic example of the importance of making simple interpersonal communications available in Africa. The instant and broad market uptake it received when made available confirmed that the killer app was the simple ability to interactively communicate. With no learning curve and no marketing, the first mobile texting service in Africa changed the course of Zimbabwe's future. The first amendment of the American Constitution ensures free speech. This comes from a very similar provision of the British Constitution that mandates freedom of expression for all British citizens. Many former British colonies, including Zimbabwe, have this provision in their national constitutions. The history of the explosion of the Zimbabwean texting platform on its then brand new mobile service, launched by Econet Wireless in Zimbabwe in the late 1990s, bears this out. For his decades as the leader of Zimbabwe since its independence from Britain

[6] Richard Close, CEO Chrysalis Campaign, Inc.; State University of New York, Empire State College, Graduate School Adult Learning; http://globallearningframework.ning.com.

in 1980, and since his later election as President in the late 1980s, Robert Mugabe managed to tightly control legislation, commerce, and expansion of his powers whenever he deemed necessary. However, once he lost the long and highly visible battle in the courts which allowed the first mobile operator, Econet, to be licensed, Mugabe's ability to push through his agenda diminished.[7] Even before Econet was able to figure out how to bill for texting, the mobile phones were used across the country to bring about the first defeat of Mr. Mugabe's then latest attempt to expand his power. The simple national texts read, "No forex, no fuel, vote no." And "no" it was, for the first time, with most of the country collaborating with itself and affecting national policy based on personal communications.[8]

Expanding this concept to interactive communications using the Internet, and adding even audio broadcast as did the Hudson study, takes this national/community conversation to the next level, where participant subscribers can share information, new approaches, and results with each other. Naturally adding video expands the possibilities and the applications in the financial, merchandising, medical, educational, agricultural, and all other sectors. But the underlying "product" is not an individual application but the empowerment of a people anxious to communicate with each other and to do so personally and commercially using ICT, the Internet, mobile phones, and as broad a band as investment allows, to connect users to each other interactively.

[7] In a civil case brought before the Supreme Court of Zimbabwe and subsequently as a constitutional case before the Supreme Court of Zimbabwe sitting as a Constitutional Court, Judith O'Neill as telecoms legal strategist and the Zimbabwean and South African team she put together of constitutional scholars won an unusual and highly precedential case in the late 1990s for the company that became Econet Wireless. The case was a constitutional challenge to the monopoly of the Zimbabwean Government-owned telephone company (ZPTT) over mobile services, which the company did not provide. While ZPTT's penetration was considerably less than 1 % on its landline (meaning less than 1 % of Zimbabweans had access to communications at all), and it did not provide a mobile service, it nevertheless relied on its legislated monopoly over all telecommunications and its right as regulator to refuse to license anyone else to provide mobile service. The Constitutional Court, relying on South African and US legal precedent, ruled that access to communications was fundamental to the entitlement to freedom of expression in the Zimbabwean Constitution. It further ruled that the constitutional exception to this provision that allowed ZPTT's monopoly over communications to not violate this provision was itself subject to exception if, as noted in the Constitution, it interfered with the principles of a democracy. The Supreme Court of Zimbabwe sitting as a Constitutional Court ruled that the ZPTT's refusal to license Econet to provide a mobile service within, into, and from Zimbabwe violated the freedom of expression provision of the Constitution and judicially mandated the licensing of Econet as Zimbabwe's first mobile carrier.

[8] Due to failing economic policies of Mr. Mugabe's Government, foreign exchange had dried up in Zimbabwe and fuel was at a shortage and caused the cancellation of many international flights during which time Mr. Mugabe sought to expand his powers so as to keep control of his programs in the face of increasing civil discontent. No forex, no fuel, vote no was the slogan that went viral over the SMS platform of Econet's mobile phones that caused the first defeat of Mr. Mugabe's agenda since his inauguration.

The Road Ahead

Similarly to the notion of connection and interpersonal communication in Africa, Close uses something called the Global Learning Framework, a logical progression of the learning process, from occurrence of the question or the inquiry to the publication and sharing of the answer. The process, while linked by Close to the functioning of the Web and of Web learning processes which feeds his theory of collaborative communication as the "killer app," is another example of the power and importance of human interactive communications facilitated by the Internet and expanded in capability and relevance by broadband and next-gen broadband.

Community organizations, entrepreneurial "clubs," trade organizations, and other proactive organizations have been developed in other countries to foster private investment and to be proactive about facilitating such investment. Local business people, with local or foreign partners, have put together model projects which they assess for attractiveness for multilateral funding, governmental support, or straight private investment based on their own assessment of their product and their market for community broadband services. The Interactive Voice Response (IVR) farms discussed in Dr. Hudson's chapter is a perfect example of how such a project might start: a business model is developed for its sustainability and how it may be advanced to video interaction, eventually enabling next-gen broadband as a video platform in Africa.

Entrepreneurial groups may be active not only in suggesting new avenues for private investment through laws and regulations, where necessary, but also in creating model projects, facilitated by government, in a transparent, cooperative manner. The solution for Africa lies in both a strong and active public as well as private sector.

Conclusion

The only certainty in Africa is one I have distilled over 20 years working in telecoms there—Africans are aware of, hungry for, and eager to put into use new technologies that allow ever more inclusive, substantive, content-laden, and affordable communications.

While Africa's telecommunications infrastructure situation may in many ways be unique, African stakeholders can still learn from and derive efficiencies from the experiences of others. The types of challenges that Africa faces, at the beginning of its journey to next-gen broadband, require creative solutions for the deployment of skills, funding, governmental support, and regulatory incentives and perhaps above all may depend on harnessing local interest. The Close model, and a similar model proposed by Bruce Lincoln, a Fellow at CITI and broadband entrepreneur for economically disadvantaged communities in the USA, combine urban poor or rural real estate investment with broadband build-out. This allows for private investment in multiple-use real estate in the target areas, into which is built mobile or fixed Internet capability which can be shared in traditional office or other commercial spaces, a centralized Internet café, learning center, and/or broadband office for rent by local enterprise. As Close suggests this may be successfully replicated in Africa. Using the same creativity, the Hudson farm study puts this into practice by combining ICTs with radio broadcast to create interactive farming communications among the farming community. In the Hudson study, the shared-space Internet facility combines commercial office use with educational facilities all connected to the Internet.

The paramount requirement for commercial success of these and other such community projects may be government's commitment to encourage and transparently support private investment in such projects. This may be done in many ways. These include (1) supporting potential investor's bids to multilaterals for funding; (2) opening the universal access fund and supportive regulations to be used for discrete investments in such projects; (3) providing tax incentives to private investors for the creation of certain facilities that will bring jobs and expanded connectivity to disenfranchised segments of society; (4) making land use laws flexible to accommodate shared-use Internet-ready facilities; and (5) creating a cyber park land area into which both government and private sector invest and whose facilities may be shared by operators, telecom incubator entrepreneurs, and universities. These are examples only and should be matched by whatever is the right formula for a country to communicate to potential entrepreneurs, investors, and multilaterals that the country is a supportive, transparent environment in which to launch next-generation broadband.

Chapter 2
Media Entertainment as a Development Strategy

Eli Noam

Rarely need one justify a topic as much as online entertainment for Africa. There is a lot of headshaking and muttering that it is not really important for Africans to watch TV shows on the Internet and that in any event their basic networks are too far behind to make this a realistic issue.

One can respond to such critique by being defensive: online TV entertainment is happening anyway in Africa, even though this might be unknown to Northern observers, and for better or worse we need to be prepared for it. But the alternative is to rejoice. Instead of wringing our hands about a frivolous distraction we should in fact positively embrace this evolution as a great opportunity for development.

So first, yes, it is happening. TV media are progressing rapidly in richer as well as poorer countries. Media firms and ICT such as BBC, NHK, Netflix, Apple, Google, and YouTube have been pushing the envelope. There are clouds, tablets, and apps. There are virtual reality, virtual worlds, and multiplayer video games. TV is changing before our eyes. It is moving into its next stage, the online stage, after the first stages, those of broadcasting, of multichannel (satellite and cable), and of digital broadcasting.

Online TV is not simply more TV on a different platform. High-speed broadband enables interactive and participatory forms of TV media, with great sharpness of resolution, wall-sized screens, and large program libraries. This is about to happen.

And one should not expect that the urban elites in poor countries will sit by and watch one or two public and commercial channels while their peers in other countries watch TV as a 2-way, 3-D, 4 K, 5.1 surround sound, 6-foot-screen medium.

Is this TV important for Africa? An Internet with video speed seems an unaffordable luxury, a cruel dream. But some of the same people said something similar, 10 years ago, about the notion of a mobile phone being in the pockets of a

E. Noam (✉)
Columbia Institute for Tele-Information (CITI), Columbia University Business School, Columbia University, New York, NY, USA
e-mail: enoam2@yahoo.com

J. O'Neill et al. (eds.), *Broadband as a Video Platform*, The Economics of Information, Communication, and Entertainment, DOI 10.1007/978-3-319-03617-5_2,

majority of Africans. And yet that's what happened. Indeed, this is not only a logical but compelling conclusion when one appreciates that before the explosion of mobile phones in Africa, televisions significantly out-penetrated fixed telephony by huge margins demonstrating the continental demand for TV. Couple the ubiquitous mobile phone with the cultural demand for TV and the relevance of online entertainment in Africa is inescapable.

I read a newspaper story report of a riot by Zambian school girls over restrictions on their cell phone use during class time. The girls were identified as "mostly from poor families." What does this tell us? First that Zambian girls are a spirited bunch, which bodes well for democracy and for equality. But it also tells us something about the essentiality of mobile communications now. Ten years ago nobody had it. But today, take it away for a few hours and you start a riot.

Mobile communications were not planned by governments as a priority. It was not declared by the World Bank as a development goal with a lot of money to back it up. But it grew and grew anyway, and due to its usefulness it created its own demand, which was supplied by multiple carriers.

Mobility was the killer app for narrowband telecom connectivity. And now, it is used for much more than voice calls to one's mother or customer. It quickly became the SMS platform even before carriers could figure out how to bill for texting. Then it becomes the mail platform, though still lagging behind texting due to pricing regulation and handset technology. In more recent years it created a more convenient and user-friendly financial facility for the growing middle and established upper classes and a whole new banking system where none existed before for the millions of previously unbanked Africans. And that's just the beginning.

In the same way, the next generation of broadband will progress. People need a killer app to get broadband and pay for it. One can talk about e-government, e-education, and e-health all day long, but that's like saying that spinach is good for you, so kids will want to eat a lot of it. People need something that excites them enough so that they will pay.

That killer app is entertainment: video, TV, films, sports, and music. For that, people will pay, save, and invest. People consume entertainment voraciously. They watch typically 4 h a day of TV, high bit rate information. 3.5 Mb per second. 4 h a day. They are often willing to pay for it. Directly or through advertising that they pay for in their product prices. And that demand will generate the business incentives to upgrade the infrastructure, first in mobile wireless and then in wireline core network as it migrates closer to the users in order to handle the loads.

Take Zambia. If of its 13 million people 10 % will find the content compelling enough to pay 8 cents per hour of entertainment, that's 10 dollars a month, and it generates about 200 million dollars per year. This creates a funding mechanism.

Thus, entertainment is enabling the economic foundation for networks. And the other applications can then piggyback on it.

Everyone around the world is promoting broadband as the base for the information society: applications such as energy, environment, health, and government.

The merits of these programs are identified, the impacts are measured, and then a pitch is made for supportive government and international funding and policy.

But let me suggest that we need to change these dynamics that are centered around government and international investments, around a supply-driven focus. Instead, the future of broadband will be demand driven: just as these school girls and other users have been driving the demand for narrowband network services. Governments may have helped, but the global prevalence of these trends shows that particular government policies do not make much of a difference. A demand explosion was happening anyway, and all that governments could do was deal with problems of equality and gaps, and otherwise just take the credit.

People will then be able to use the networks in many ways, representing the needs and desires of a broad cross section of the society—students, doctors, teachers, creators, retailers, banks, etc. And therefore, by unleashing this demand, a video policy is a development policy.

Of course, no gain comes without pain. In particular, the media system of countries becomes challenged, and this is typically a politically highly sensitive area. Public broadcasters lose some of their audiences, though they might gain others. National media producers have to contend with a greater foreign competition, though they may also have opportunities to export themselves.

Will the content be local, or is it necessarily global? The answer is both. There will be international, high-value entertainment of the Hollywood type. But there will also be content from nearby countries, providing for regions that share a similar culture. And thirdly, there will be local content in local languages from national public broadcasters, private media firms, and many small independents.

Media regulators will have greater difficulties to control such an emerging complex media system. Their policy goals deserve respect and do not necessarily change just because we use digital packets and IP instead of analog. But it would be a very expensive mistake to let the slow-moving media regulatory system keep network evolution hostage to a slowly reforming media policy.

Importantly, encouraging entertainment applications does not mean frivolous use. It does not mean giving priority to investments so that people can watch reruns of "Who Wants to be a Millionaire."

It means transforming network development from a supply-driven, network-driven, government-driven emphasis to an orientation that is user driven, applications driven, and demand driven. And which, to boot, will cost the governments very little, outside of the high-cost, low-density areas, to create a national infrastructure.

So media policy will become development policy, and entertainment policy will be infrastructure policy.

Chapter 3
Let Them Eat Movies: (How) Will Next-Generation Broadband Diffuse Through Africa?

Indra de Lanerolle, Alison Gillwald, Christoph Stork, and Enrico Calandro

"Wireless mobile services for the delivery of broadband provides false hope for the future and promotes inefficient use of resources in the present", Eli Noam argued in Lusaka,[1] one of the least connected capitals in the world. Applying his 2011 paper, *Let Them Eat Cellphones: why mobile wireless is no solution for broadband,* in which he rejects proposals in the United States for the deployment of mobile wireless services to meet rural broadband needs to Africa, he argued that the cost of mobile wireless deployment was only cheaper in the short term. The high costs associated with matching constantly growing demand with limited supply of spectrum, he argued, meant that mobile did not enjoy the longer term economies of scale that fixed-line investments did. Noam also cautioned against underestimating the power of video, to which he attributed broadband take up in the United States. He challenged the notion that communities forced to access video on mobile wireless networks would simply accept poorer quality services than their urban or wealthier counterparts, who were able to receive the array of high-quality digital services on offer.

There are compelling aspects to Noam's arguments. "Speed" or bandwidth is an important aspect of Internet functionality. The Internet is a global, general-purpose technology network that enables delivery of a range of communication, information and entertainment services. The designers and providers of these services—rarely African—must make decisions based on their assumptions about the bandwidth that their users have access to. As access speeds increase globally, the bandwidth demands for services increases. Voice over IP and video streaming services, for example, do not work well, or at all over slow connections. Facebook, originally largely a text-based service, now prioritises photos in its news feeds.

[1] Noam gave a keynote address at a conference in 2012 held in Lusaka, Zambia entitled: Next Generation Broadband as a Video Platform—Strategies for Africa. Two of the authors presented papers at this conference on which this chapter is based in part.

I. de Lanerolle (✉)
University of Witwatersrand, Johannesburg, South Africa
e-mail: indra@delanerolle.net

J. O'Neill et al. (eds.), *Broadband as a Video Platform*, The Economics of Information, Communication, and Entertainment, DOI 10.1007/978-3-319-03617-5_3,

Can and will such infrastructure get built on the African continent? If not will Africa simply fall further and further behind the rest of the world? Or will it, as it has in the past, modify, adapt to and occasionally even innovate with what is available?

Over the last century only two kinds of network infrastructure, both communication and both wireless, have succeeded over most of the continent of sub-Saharan Africa. The first is the group of national terrestrial broadcasting networks—radio and television. The second is the group of cellular wireless mobile phone networks. Many other network infrastructure exists—from road and rail to fixed-line telephones and even electricity grids—but almost invariably they have failed to reach scale and widespread use. As Yepes et al. (2008) have shown (Table 3.1), in such cases, African countries have achieved far lower levels of penetration than other parts of the world with similar levels of income.

Noam (2008, 2011) has made four important and related points about next-generation broadband[2] (NGB) and its relevance to Africa. First of these is a technical argument that NGB is required for only certain applications—those that require real-time or fast delivery of new forms of very rich video content—richer than current broadcast television standards including high definition (HD) (Noam 2008). Second he makes a related social and economic argument (Noam 2008) that this video content will be primarily entertainment content—as he put it in Lusaka: "entertainment will be the killer app" for NGB. Thirdly, he has made a further technical argument that NGB can only be delivered by fixed connections since wireless technologies are—and will be—incapable of delivering the required throughput efficiently (Noam 2011). Lastly, he has proposed that these arguments do not apply only to developed countries, such as the United States, but will also apply in Africa.

This chapter explores Noam's contention, based on these claims, that fast broadband or NGB must be the next important network infrastructure to come to Africa. It examines Noam's argument that demand for video entertainment which has driven the demand for faster broadband in the United States and other rich countries will drive the demand in Africa. It further considers the implications of his argument that fibre is the best and possibly the only means of delivering the very high bandwidths envisaged as NGB.[3] In considering the application of his arguments in an African context, we address two questions.

[2] There is some variation in the use of the term NGB. Noam (2008) defines it as bandwidth of 1 Gbps. The UK Government on the other hand recently defined it as 250 Mbps. As the term implies, what is "next" depends on where you start. This issue is discussed in the conclusion.

[3] We note Noam's arguments concerning the technical limitations of wireless transmission. However, "forever" is a long time, and while Noam (2008) does allow for future improvements in compression technologies for example, it seems to be possible at least that his estimates of these improvements could in principle be outperformed through technological innovations not yet foreseen. Samsung recently announced that they had successfully transmitted data at over 1 Gbps over a 2 km distance (Samsung 2013). Further, currently 3G and LTE networks can offer better throughput than the poor-quality ADSL available in South Africa.

Table 3.1 Network infrastructure in Africa

	Low income			Lower mid income			Upper mid income		
Sector	Africa	Other	Ratio	Africa	Other	Ratio	Africa	Other	Ratio
Density of paved road in km/1,000 km sq (2001)	31	134	4.3	94	141	1.5	238	781	3.3
Fixed-line density: subscriptions per 1,000 people (2004)	10	78	7.7	106	131	1.2	120	274	2.3
% of households with access to electricity (2004)	16	41	2.6	35	80	2.3	28	95	3.4
Density of mobile phone subscriptions per 1,000 people (2004)	55	86	1.6	201	298	1.5	422	554	1.3

Adapted from Yepes et al. (2008) p. 7

- On the supply side: In what ways is Africa different from the United States or the rest of the world in the capacity of firms or governments to deliver the infrastructure required (e.g. fibre or cable) to African homes?
- On the demand side: In what ways is Africa[4] different from the United States in particular or the rest of the world in general in its demand for entertainment, specifically video entertainment (and are those differences significant in how they might affect the demand for NGB)?

Forecasting the adoption of new technologies in Africa is notoriously inaccurate. Mobile services were introduced as a limited elite service to provide complementary mobile services for those with fixed lines at work and at home. Early business models forecast subscriber numbers at little of a quarter of what they achieved in the first 5 years in South Africa for example.[5] On the other hand nothing is inevitable. The skeletons of perfectly feasible services litter the continent.

Rather than presenting an alternative scenario on what the broadband landscape will or should look like in the future, the first section of this chapter examines some of the assumptions that underpin Noam's argument in the African context. Having considered some of the supply-side issues relating to existing networks and services, the institutional capacity required to create efficient markets or state delivery of broadband and the impacts of this on the cost of communication, it examines

[4] "Africa" is a broad term. Here we are concerned with sub-Saharan Africa, which in terms of ICT infrastructure differs substantially from North Africa. However even within sub-Saharan Africa, as Table 3.2 indicates, the development of ICT infrastructure varies greatly and the conditions (economic, political, geographic and social) in which such infrastructures develop vary greatly. So the generalisations offered are exactly that.

[5] "Vodacom's initial growth projections catered for 250,000 subscribers within 10 years. We have now exceeded three million during the year 2000." Vodacom quoted in African Wireless. http://www.africanwireless.com/vodacom_history.htm accessed 17th June, 2013.

some of the demand-side issues in the second section. This second section examines current demand for entertainment and information services and the drivers of broadband Internet from a nationally representative 12-country African household and individual user survey conducted during 2011–2012 (RIA Survey 2011/2012; see Appendix for details).

The final section considers the implications of Noam's contention for Africa in the context of existing infrastructure and institutional constraints and explores the likelihood of this coming to pass in the short to medium term.

Communications Infrastructure

While a number of the assumptions that underpin Noam's arguments may be true for the United States they are not necessarily correct for Africa. Making his case for fixed high-speed broadband deployment rather than wireless technologies, Noam points out that in the United States most rural households already are not dependent on wireless for broadband. The majority of rural homes are passed by cable TV or phone connection which enables much faster speeds and are likely to be upgraded given past trends. The relatively few homes that are not connected to any communications network "can be served by fixed wireless, provided by entrepreneurial wireless Internet service providers (WISPs) and their high-speed directional microwave service, without such a tiny tail wagging the rest of the country" (2011:481).

Infrastructure to Homes

The situation in African households is starkly different. The availability of electricity continues to be a major determinant of which communication modes are available and present at the household level, especially television, and at a more basic level, in seven countries surveyed, water is piped into less than one in ten households (Table 3.2).

Aligned to conventional wisdom on media access use in Africa, the 2011/2012 RIA Survey confirmed that in almost all the 11 African countries under investigation, the radio is the most widely owned electronic appliance for information at a household level compared to TV and the Internet, with the exceptions of Cameroon and South Africa, where TV is now the main electronic information means at a household level. Radio penetration at a household level across these 11 countries reached 40 %, and in countries such as Ghana, Kenya, Namibia, Rwanda and Uganda it is above 70 %. Rwanda and Uganda have a very low level of electrification at a household level and almost no Internet connectivity at the household level. In these two poor countries, radio still represents the only affordable electronic appliance for information.

Table 3.2 Households with working ICT appliances

	Main electricity grid (%)	Water piped into the house (%)	Fridge (%)	Radio (%)	TV (%)	Satellite decoder or cable (%)	Computer (%)	Pay TV (%)
Uganda	13.4	1.5	5.5	76.6	12.9	1.5	2.2	2.3
Rwanda	15.6	1.1	3.0	72.4	9.0	1.7	2.0	2.0
Ethiopia	18.1	0.3	3.1	40.7	10.0	3.0	0.7	0.0
Tanzania	19.4	0.6	8.5	63.1	18.3	3.2	1.6	2.5
Namibia	41.8	32.4	40.6	72.0	40.6	17.7	14.7	20.9
Nigeria	58.4	2.2	24.3	69.5	53.0	12.9	6.6	10.2
Kenya	60.1	14.6	14.1	80.6	54.4	6.2	12.7	3.2
Botswana	60.1	22.7	53.6	66.4	59.4	43.5	15.8	12.6
Cameroon	64.5	6.2	14.8	33.9	44.3	21.1	8.6	13.1
Ghana	73.0	7.8	36.7	71.8	54.1	8.0	8.5	4.4
South Africa	89.2	36.3	74.2	62.3	78.2	34.9	24.5	29.7

Source: RIA 2011/12 Survey

In a few countries such as Ghana, South Africa and Nigeria, radio ownership at a household level is decreasing compared to 2007/2008 penetration levels. These countries have a higher GDP per capita than the other countries under investigation, and therefore households can afford alternative information means such as TV.

Ownership of a working TV at a household level has grown in almost all the 11 countries in 2011/2012. It is still equal or below 40 % in Namibia, Rwanda, Tanzania, Uganda and Ethiopia. The main reason given for a low level of TV penetration in these countries is the lack of electricity at home which reflects the lower levels of electrification at a household level in these countries.

Internet connectivity at a household level is negligible in the majority of these countries, at below 4 %, except in Botswana, Kenya, South Africa and Namibia. However, the RIA 2011/2012 survey shows that penetration levels of Internet at home are growing at a fast pace, though off a very low base, with changes between 2007/2008 and 2011/2012 higher than 8 % in Botswana, Kenya, Namibia and South Africa. This low level of Internet connections to the home follows the failure of fixed-line networks to diffuse widely in Africa. Of the countries included in the RIA Survey only South Africa and Cameroon have achieved household penetration rates of over 10 % (Table 3.3).

Most Internet research in Africa therefore is still concerned with questions of diffusion of the current networks, broadband or not, rather than imagining if, or under what conditions, new generations of technology may be rolled out.

Mobile Networks and Devices

Noam also challenges us to look beyond mobile. This is difficult to do in the African context. In contrast to this rather dismal picture of infrastructure to the home, the

Table 3.3 Internet access via mobile phones

	Per 100 adult population (15+)					
			Among mobile phone owners			
	Internet users	Mobile phone owners	Internet-capable mobile phone owners	Using mobile to browse the Internet	Computer (desktop or laptop) owners	Fixed-line penetration per 100 households
Ethiopia	2.7	18.3	6.5	5.1	0.5	4
Tanzania	3.5	35.8	19.2	5.2	0.9	0.4
Rwanda	6.0	24.4	19.1	14.9	1.8	0.2
Uganda	7.9	46.7	14.9	7.7	2.0	1.5
Ghana	12.7	59.5	28.5	13.4	7.4	1.8
Cameroon	14.1	44.5	14.9	8.1	5.7	2.2
Namibia	16.2	56.1	30.7	23.8	9.3	11.5
Nigeria	18.4	46.7	22.7	16	5.0	0.3
Kenya	26.3	74	32.3	25.3	11.0	0.6
Botswana	29	80	29.5	22.8	NA	15
South Africa	33.7	84.2	51.0	27.6	17.6	18

Source: RIA household and individual survey 2011/2012

success story of the Internet in Africa is mobile, both mobile devices and mobile wireless networks, as many others have pointed out (see for example, Stork et al. 2013; Williams et al. 2011). According to the RIA Survey mobile phone adoption rates in the countries surveyed vary from 18 to 84 per 100 (Table 3.3).

This user base has in turn allowed significant growth in Internet adoption (Table 3.4). South Africa has the highest Internet penetration rate among all the countries under investigation, with 33.7 % of the population aged 15 years or older using the Internet. Kenya, a low-income country, almost doubled its level of adoption from 15.0 to 26.3 % over 4 years as did Namibia where Internet users increased from 8.8 to 16.2 %. Botswana had the highest growth of Internet users compared to 2007/2008. The number of Internet users also grew significantly in Uganda, Rwanda and Tanzania compared to 2007/2008, when a very low percentage of the population was connected to the Internet. Ethiopia has the lowest level of Internet penetration with less than 3 % of the population using the Internet. This is likely the outcome of policies that have prevented the kind of competition that drives take up in other jurisdictions. The Ethiopian market remains structured around a monopoly operator, where a single operator provides fixed, mobile, and Internet services and maintains the international voice and data gateway (Adam 2010).

While the first wave of Internet adoption rode on the back of desktop computers for the elite at home and for others at the work place, schools and universities or public access facilities such as Internet cafes, the second wave sweeping across Africa is through the use of mobile phones. Although Internet penetration in most countries is still very low, more than 70 % of Ugandan and 67 % of Ethiopian

Table 3.4 Individual Internet use

	15+ that use the Internet			Where the Internet was first used		Where did you use the Internet in the last 12 months?				
	2007/2008 (%)	2011/2012 (%)	Diff. (%)	Computer (%)	Mobile phone (%)	Mobile phone (%)	Work (%)	Place of education (%)	Another person's home (%)	Internet cafe (%)
South Africa	15.0	33.7	18.7	65.1	34.9	70.6	35.8	20.9	14.3	32.4
Botswana	5.8	29.0	23.2	70.6	29.4	64.1	51.1	32.2	43.7	58.3
Kenya	15.0	26.3	11.3	68.9	31.1	77.8	31.4	38.8	38.9	72.4
Nigeria		18.4		45.2	54.8	74.9	29.3	19.6	30.3	45.1
Namibia	8.8	16.2	7.4	50.1	49.9	87.3	48.4	36.0	32.6	22.5
Cameroon	13.0	14.1	1.1	82.1	17.9	29.7	9.8	20.1	18.7	80.0
Ghana	5.6	12.7	7.1	70.5	29.5	61.2	34.6	50.9	34.5	84.7
Uganda	2.4	7.9	5.5	28.2	71.8	81.3	55.0	51.2	54.0	74.0
Rwanda	2.0	6.0	4	70.8	29.2	70.9	52.1	30.7	24.9	50.2
Tanzania	2.2	3.5	1.3	45.8	54.2	74.7	44.6	24.4	23.9	62.8
Ethiopia	0.7	2.7	2	33.3	66.7	80.9	17.4	20.9	3.5	42.2

Source: RIA 2011/2012 Surveys

Internet users first used the Internet on a mobile phone. In Tanzania, Namibia and Nigeria about half of the population first used the Internet on a mobile phone (Stork et al. 2013).

The growing importance of the mobile phone to access the Internet is also demonstrated by the responses to the question "Where did you use the Internet in the last 12 months?" Either complementarily to computer access or exclusively, above 80 % of Internet users in Namibia, Uganda and Ethiopia access the Internet via a mobile phone. In South Africa, Kenya, Nigeria, Tanzania and Rwanda the Internet is accessed through a mobile phone by 70 % of Internet users. In most countries, besides Cameroon and Ghana, the mobile phone has overtaken the Internet cafe, historically the most common way to access the Internet.

The increase in usage intensity is also a positive development. Daily usage rose over the past 4 years compared to once-a-week and once-a-month usage, particularly in Rwanda and Ethiopia (Fig. 3.1).

Linked to the increase in the frequency of use, the figure above depicts the magnitude of the new wave of Internet users. In Ethiopia, almost half of the Internet users started browsing the Web during the last year. By contrast, the share of early Internet adopters (share of Internet users that started using the Internet 4 years ago or earlier) is larger in Botswana, Rwanda, South Africa and Ghana.

While Namibia reflects early adopters already forming a wider base, Rwanda's continued low rate of adoption of mobile Internet may have something to do with the state-sponsored programme to promote low-cost computer uptake, together with limited 3G or better mobile data access, at a time that the market was meeting pent-up demand to access the Internet through mobile access.

This evidence shows that the infrastructure situation on the ground in Africa is exactly the inverse of the situation Noam describes above in the United States. Almost all rural (and indeed many urban) homes are not connected to any fixed-line communications network (cable or phone, copper or fibre). The primary means by which people communicate and the primary means by which Internet users connect are via mobile phones and mobile wireless networks.

So if Noam is correct that NGB can only be efficiently delivered by fixed fibre networks then this would require that the limited reach of fixed infrastructure identified above, beyond the main backbone, would be overcome.

Demand for Entertainment and Video Services

In the United States and elsewhere, the most important applications driving demand and supply of fast broadband appear to be those that compete with or substitute for broadcast (free or pay) television and the entertainment video market. A recent study by IDC for example shows that download video content in the United States exceeds uploading by more than ten to one (IDC 2012). Another study by CISCO systems estimates that video now accounts for most data transferred over the Internet (CISCO 2011). The online video download service Netflix is now the market leader in the

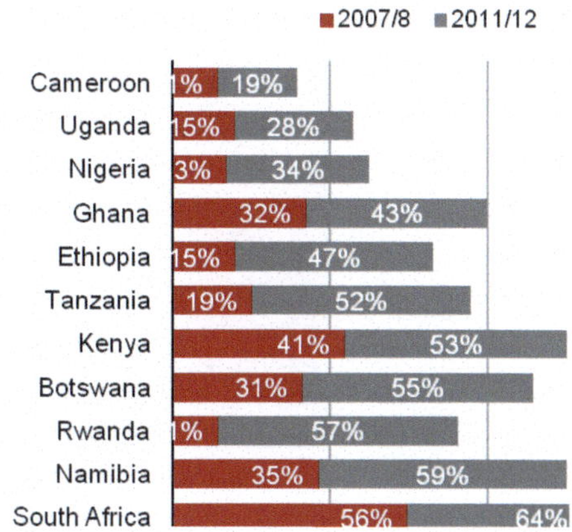

Fig. 3.1 Frequency of Internet daily usage in the last 3 months (*source:* RIA 2011/RIA 2007)

video market ahead of bricks-and-mortar businesses like Blockbuster and cable pay-per-view services like Comcast (Noam 2011). While there are many other uses of the Internet, given innovation in traditional television services (ultra HD, increasing television screen sizes, and 3D television for example) the bandwidth required to provide these services via broadband means that in the United States, these broadcast-like services may be the key driver of demand for increased bandwidth. In South Korea, the global leader in terms of broadband speeds to the home, video gaming was an important driver of early broadband adoption (Yun et al. 2002). The World Internet Project reports seven countries where gaming is more popular than video and seven countries where the reverse is the case (World Internet Project 2012). But in both cases, it is the requirement for high-quality real-time moving images that drives the demand for fast broadband. This evidence supports Noam's technical arguments concerning both the specific range of applications for which fast speeds are required and the prime place of entertainment in driving demand. But what evidence is there that this claim is true for Africa? Our research suggests that demand for Internet video in Africa differs from many other countries due to differences in the broader communications landscape.

Television in Africa

Globally, television is by far the most successful entertainment distribution system yet devised. Even amongst Internet users, in all 16 countries surveyed in the World Internet Project in 2009 and 2010, most people regarded television as important or very important as an entertainment source and in all but two cases, more so than the Internet (WIP 2012). Despite the so-called 'mobile miracle', in a number of African countries, significantly more people have access to television than to mobile phones and even those with mobile phones depend on traditional communication services—radio and television—for their informational and entertainment needs.

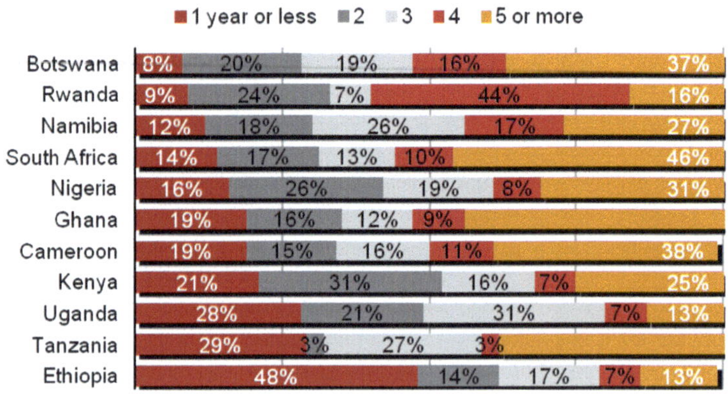

Fig. 3.2 Share of Internet users by years since they first used the Internet

There is, however, a shift to acquiring their information needs from the Internet through their mobile phones as we see below. While the RIA Survey shows that television use remains below 40 % in many countries in Africa (Fig. 3.3), this does not necessarily reflect demand. Rather, this lack of usage is often due to the absence of electrification or availability of broadcast services, which seems to be the case in Namibia, Tanzania, Uganda and Rwanda. Other than Namibia, these are also countries which have lower GDPs per capita. In fact all evidence suggests a pent-up demand for information and entertainment services.

In support of Noam's contention on entertainment driving uptake, the RIA Survey demonstrates that those with access to television rated entertainment highly. When asked what channels or programmes they like to see more in the context of digital migration, entertainment is the one that the majority of TV watchers want to see much more of (Figs. 3.4, 3.5, and 3.6).

As importantly, this research shows that television audience consume large quantities of broadcasting. Most TV viewers watch for more than 1 h a day. Only in Ethiopia and Namibia did respondents seem to watch TV irregularly and infrequently, while in the countries with bigger TV audience such as Kenya, Nigeria, South Africa and Botswana most TV viewers watch TV for 2 h and more a day (Table 3.5).

Where affordable and accessible, the preference is to watch TV at home. Only in Tanzania, Uganda and Ethiopia which are the least electrified does public TV viewing play an important role.

The main reasons for not watching TV are lack of electricity at home and the high cost of acquiring a TV set (Table 3.6). Lack of electricity was the main reason for 85 % of Ugandans and 89 % of Tanzanians that do not watch TV (Table 3.7).

This evidence supports Noam's contention that entertainment is the key to understanding demand. This echoes a study which found that despite the focus by donor and multilateral institutions on educational, health and other developmental applications in Africa (see for example, Qiang et al. 2009), that were believed to be what people at the base of the pyramid needed, what all people actually want is very

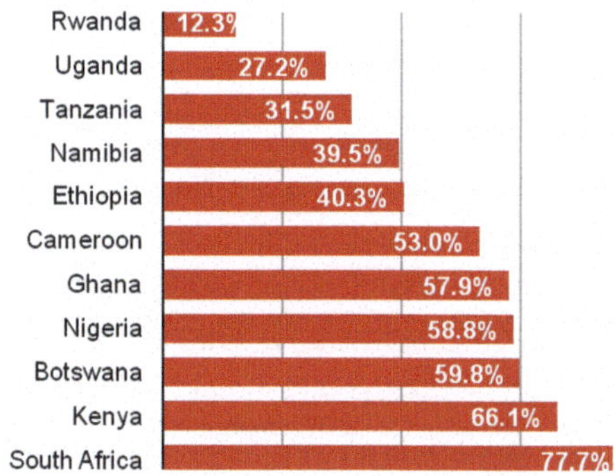

Fig. 3.3 Share of individuals watching TV (*Source:* RIA 2011/2012)

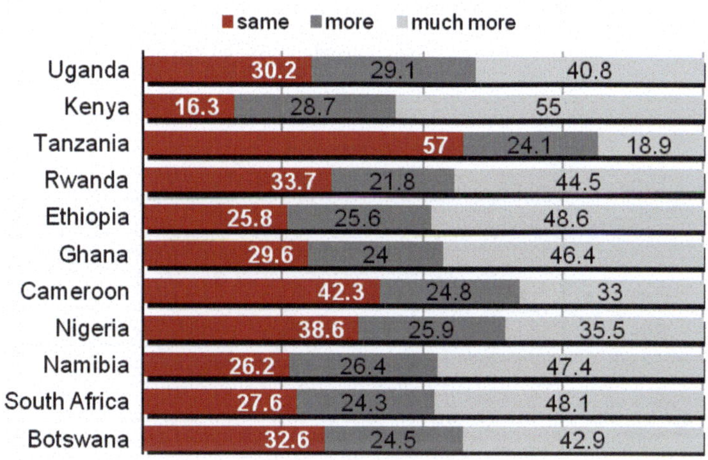

Fig. 3.4 Entertainment: If more channels were made available, which programmes would you like to see more?

similar to the things people at the top of the pyramid want, just more affordably. This includes games, social networking, music and video shorts (see also Calandro et al. 2012).

Can African Consumers Pay for Video Entertainment?

These patterns of television consumption put mobile phone use into some perspective. In South Africa for example, the median average amount spent on mobile airtime per month is around R70—less than \$8[6] which at 2012 average call rates in

[6] The exchange rate in January 2013 was approximately 9 South African rands to the US dollar.

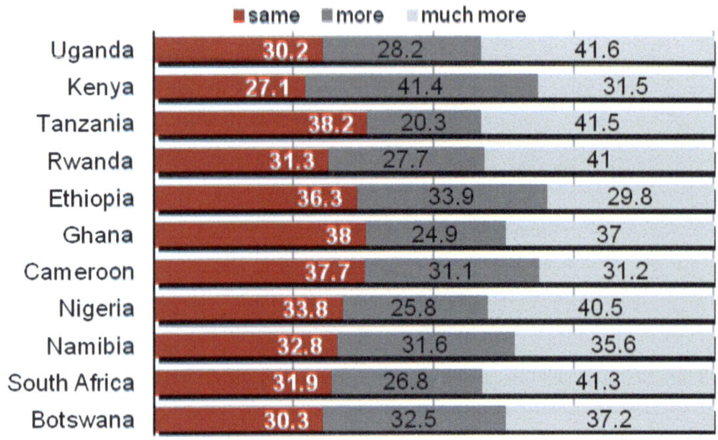

Fig. 3.5 News: If more channels were made available, which programmes would you like to see more?

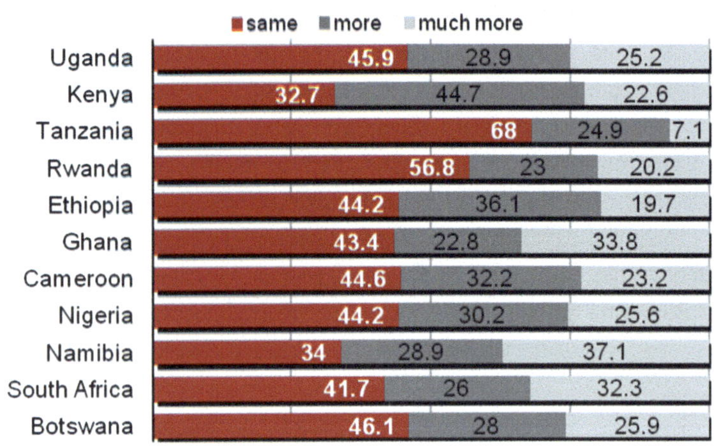

Fig. 3.6 Educational programmes: If more channels were made available, which programmes would you like to see more?

South Africa (Research ICT Africa 2012) could represent less than one 3 minute call per day.[7]

However, while African consumers may want such entertainment, the question remains as to whether economic demand is sufficient to support investment that would be required to service it via NGB fibre networks.

[7] This does not include time spent receiving calls. Based on the analysis of the RIA Household Survey, mobile phone users most commonly report receiving as many calls as they make.

Table 3.5 How many hours a day do you watch TV?

	Very irregularly (%)	Less than 1 h (%)	Between 1 and 2 h (%)	Between 2 and 4 h (%)	More than 4 h (%)
Uganda	26.2	9.0	31.2	18.5	15.1
Kenya	5.9	5.1	25.9	37.8	25.2
Tanzania	5.9	24.0	43.0	18.5	8.5
Rwanda	23.1	11.1	40.5	18.5	6.8
Ethiopia	68.2	6.9	14.4	8.3	2.1
Ghana	11.9	10.2	27.0	26.8	24.2
Cameroon	17.4	6.8	36.7	18.2	20.9
Nigeria	16.0	14.7	26.0	29.6	13.7
Namibia	31.5	6.7	16.7	17.5	27.6
South Africa	4.3	4.0	20.6	27.6	43.6
Botswana	13.0	8.1	19.0	25.3	34.7

Source: RIA 2011/2012

Table 3.6 Where do you watch TV mostly?

	At home (%)	At friends', relatives' or neighbours' home (%)	Public places (bars, community halls, TV clubs) (%)	Other (%)
Uganda	39.7	13.2	41.1	6.0
Kenya	82.5	6.5	7.6	3.5
Tanzania	53.2	22.2	17.4	7.2
Rwanda	65.0	22.8	9.8	2.4
Ethiopia	21.9	13.1	59.8	5.2
Ghana	90.1	8.0	1.5	0.4
Cameroon	85.9	9.9	3.5	0.7
Nigeria	84.7	12.1	3.1	0.2
Namibia	86.9	5.6	6.5	1.1
South Africa	97.0	2.4	0.4	0.2
Botswana	87.0	11.0	1.5	0.6

Source: RIA 2011/2012

Pay television provides some sense of who can pay for what. It has grown significantly in the last few years, especially in South Africa. However, most broadcast audiences across the continent, where they have access to television, are consuming "free-to-air" services—in other words they are not paying directly for the services they consume (Fig. 3.7). Advertising rates—the amount that marketers pay to reach this audience—provide an indicator of the economic value that each thousand viewers create for the television service. The most popular programme in South Africa for example, a soap opera called "Generations", screened in prime time on the most popular channel, had an undiscounted advertising rate of R130,000 per 30 s in January 2013 (SABC 2013). Its viewership is in the region of 6.5 million viewers. This is equivalent to a value of SA rands R0.40c or less than US $0.05c

Table 3.7 Reasons for not watching TV

	Not interested (%)	House has no electricity (%)	Cannot afford a TV set (%)	Don't have time to watch TV (%)	Cannot afford TV license (%)
Uganda	32.9	85.0	69.4	31.6	60.6
Kenya	13.5	72.5	70.6	19.7	50.8
Tanzania	8.0	89.3	87.9	34.2	83.6
Rwanda	16.5	73.8	78.6	11.3	57.6
Ethiopia	18.2	75.1	80.1	21.9	31.5
Ghana	22.7	51.1	55.0	31.0	27.8
Cameroon	25.8	60.8	73.5	33.5	62.3
Nigeria	22.9	69.6	54.6	23.1	42.8
Namibia	10.2	87.2	75.2	20.9	70.6
South Africa	17.1	35.4	59.0	12.5	44.3
Botswana	12.2	71.4	55.8	18.4	32.4

Source: RIA 2011/2012

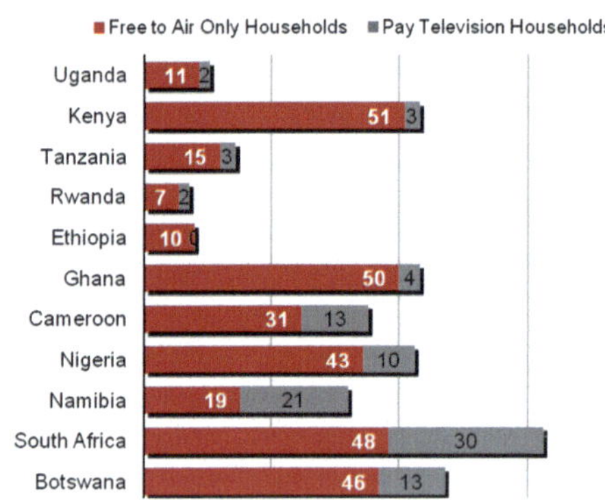

Fig. 3.7 Share of households with pay and free-to-air television

per viewer per hour for the most popular programme on television. Average prime-time television rates per hour per viewer in South Africa—the richest country with the richest broadcasting system in sub-Saharan Africa—are lower than this. We estimate, across the three most popular channels, that they are less than 25 % of this amount—around 2c (US) per hour.[8]

When considering the economic demand available to fund investment in further services we also need to take account of the fact that the current television market on the continent is highly concentrated with a relatively small number of channels available to most audience. Further services—via fibre to the home (FTTH) for example—would increase competition on the supply side and thus would be likely to put downward pressure on rates per viewer.

[8] This does not take account of discounting of published rates—a common practice in South Africa and elsewhere.

Individual Internet Use

The RIA Survey also offers evidence that Internet use in the African countries surveyed is predominantly meeting users' communication needs rather than their desire for entertainment, though entertainment dominated informational activities on a daily basis (see Table 3.8).

In all of the countries surveyed, communication services such as email or social network services were used daily by more Internet users than entertainment or information services. In eight countries, there was more daily use of one or more information services than of an entertainment service. There could be a number of reasons for this including lack of content (in comparison to what is available on other media such as television), the cost of consuming data required for viewing video content (compared to free-to-air broadcasting) and the limitations of mobile phone devices for consuming entertainment content. Given the evidence that mobile phone users ration their calls because of costs, it appears that unmet demand for communication is what is currently driving Internet diffusion in many African countries. As one young man in a village in Kenya reported to one of the authors to explain his heavy use of Facebook on his mobile phone: "its much cheaper than calling my friends".[9]

So in examining consumption of television and Internet services, we again see an inversion of the situation between technologies in the United States and in many African countries. In the United States, television is usually paid for—offering an existing market that broadband video entertainment services can compete in. In Africa the vast majority of television viewers consume only free-to-air services. In the United States, local telephone calls (on fixed-line networks at least) are often free. In Africa, voice calls are relatively expensive and only available via mobile phones. In this context, especially for low-income users, using the Internet for entertainment in Africa is expensive (compared to free television services) and using it for communication is cheap (compared to the cost of using mobile voice services).

The significant growth in Internet use in Africa over the last 4 years has not been driven by demand for online entertainment but rather by the need to communicate cheaply, and while social demand for entertainment on "traditional media" in Africa may be similar to that in rich countries, economic demand is far weaker. This leads directly to the supply-side question: Will this weaker demand be sufficient to encourage supply of NGB services, or if such services are available would they stimulate demand for online entertainment?

[9] Interview with Mwangi Nahashom, Olekasasi village, Kenya, March 2013.

Table 3.8 Internet users that perform the following activities daily

	Communication		Entertainment		Information			
	Social networking (%)	Email (%)	Gaming (%)	Download or watch movies, music, TV, radio (%)	Information: health related (%)	Find or check a fact or look up a definition (%)	From government (%)	Online newspapers, magazines, electronic books (%)
Uganda	19.4	27.9	10.6	14.0	2.4	4.0	4.2	14.6
Kenya	45.8	24.6	4.1	10.6	2.4	16.7	1.4	8.8
Tanzania	27.1	38.1	9.0	11.9	3.2	20.6	12.0	21.2
Rwanda	43.8	23.6	0.1	1.0	1.0	5.0	1.0	0.0
Ethiopia	9.3	19.0	0.0	1.2	1.0	3.9	2.1	5.4
Ghana	35.1	35.3	6.9	11.2	7.9	24.3	15.1	9.5
Cameroon	24.2	16.1	3.5	5.2	5.8	7.4	1.0	5.8
Nigeria	35.5	22.4	15.2	8.3	13.1	11.6	10.1	11.9
Namibia	31.3	28.7	10.7	12.8	9.0	30.6	7.4	19.9
South Africa	51.7	35.6	12.7	12.7	4.2	12.0	4.9	10.7
Botswana	41.4	40.1	8.1	8.9	3.6	13.9	2.8	16.5

Source RIA 2012/2012 Survey

Current and Next-Generation Networks

Any investment in the supply of new broadband infrastructure in Africa will take place in a context of the current state of the Internet network infrastructure. We have highlighted the low levels of fixed-line connectivity. But it is the limited capacity of the network infrastructure in general that gives an indication of the further challenges any next-generation investment faces.

International Connectivity and National Backbones

Following the landing of international undersea cables (Fig. 3.8), the astronomical input costs of international bandwidth, which for example used to make up as much as 80 % of ISP charges in South Africa, have plummeted in the last 3 years in many countries, particularly those with coastal landing points.

However, the lack of International Exchange Points (IXPs) in many African countries and the lack of sufficient backhaul connections, together with high transit charges between countries, have limited the benefits of these cables, especially for landlocked countries. National backbone infrastructure is also weak and very unevenly distributed with the bulk of it in South Africa where almost the entire network is owned by one firm—Telkom SA (Fig. 3.9).

The Last Mile: ADSL, LTE and DTT

Despite the gains in mobile service, Noam is right that universal access to the full range of communication services and quality of service remain crucial policy issues. The slow deployment of fixed broadband service and the high cost of this service mean that mobile broadband services have rapidly become the primary form of broadband access over the last 5 years. In Africa, mobile broadband is not a complementary service to fixed broadband as it is in mature Northern economies. While significant, mobile broadband take up was still constrained by the requirement of a computer into which wireless dongles are plugged. The availability of more feature phones and smartphones means that Africans are accessing the Internet for the first time from their mobile phones in significant numbers. The immediate relief provided by wireless and mobile services to bandwidth-starved Africans, however, has been stymied by the lack of access to next-generation spectrum (LTE). This is the result of the delays in the migration of analogue terrestrial broadcasting to digital.

At present FTTx (fibre to the home, the curb or business) is almost non-existent across the continent with minor exceptions in a very small number of cities—for example in Nairobi in Kenya and Lagos in Nigeria. While a small number of

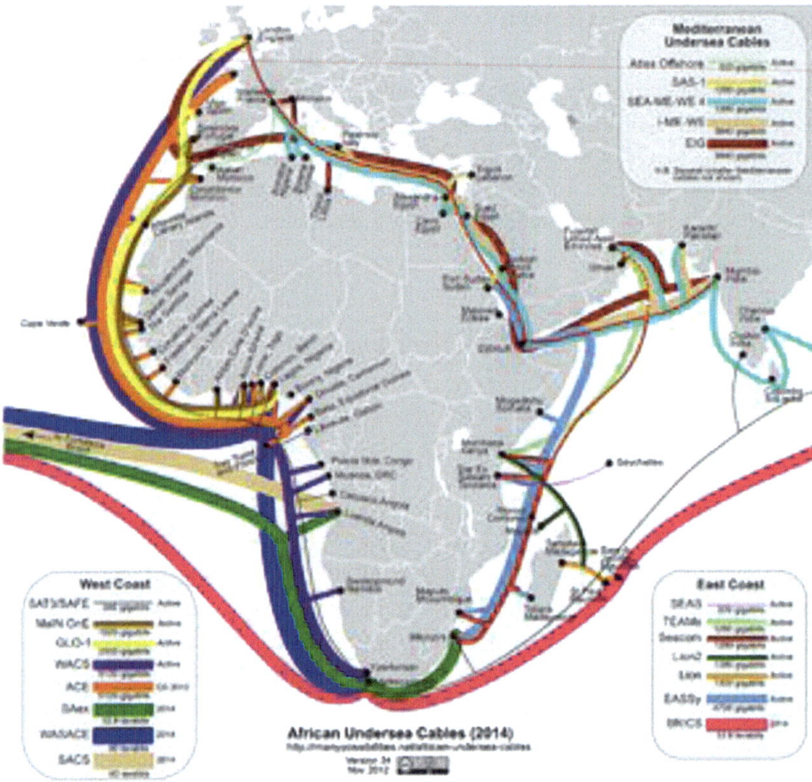

Fig. 3.8 International undersea cables in Africa. *Source:* Steve Song, http://manypossibilities.net/african-undersea-cables/

firms—largely in Southern Africa and east Africa—are investing in new infrastructure, levels of investment are still far from adequate and far from sufficiently spread to provide the basis of an NGB network or indeed of any broadband network to homes across the continent.

ADSL is still to be widely deployed, even in countries with higher levels of fixed-line networks. And where it has been deployed, it is often expensive. LTE, the fastest commercially available mobile wireless broadband technology, has not been rolled out widely, in part because of delays in "digital migration" in Africa—the replacement of analogue broadcast services with digital terrestrial services (DTT), which would make valuable spectrum available.

There are many issues facing digital broadcasting in a continent with limited access to "big" entertainment, a wide free television following, uneven levels of literacy and low household incomes. While the development of a sufficient amount of relevant content to drive the uptake of multiple channel broadcasting services is arguably a big challenge, even a limited number of alternative channels, particularly with premier sports and film rights, may present an attractive source of

Fig. 3.9 Fibre backbone networks in Africa. *Source:* http://afterfibre.net

information and communication to consumers. In some countries, the opportunity to have their first non-state-owned services might also be a spur to adoption. A multichannel environment supported by the state, would also be able to deliver formal education support services and informational services such as health, small business development, general citizen or parliamentary programming. Although people will not purchase new services in order to get e-government services, focus group research done on the Base of the Pyramid in South Africa in 2011, in a mobile phone environment, indicates that when health or e-government applications were preloaded on their devices people used them (Calandro et al. 2012). Likewise in several of the countries in the RIA Survey in 2012 education was still prioritised as a reason for watching television.

In this way freeing up high-demand spectrum for 4th Generation mobile services and the migration of analogue to digital broadcasting may meet the information and entertainment needs of African consumers who are becoming dependent on mobile communications and increasingly on high-speed mobile services, at least in the near term.

Quality of Service

Both for ADSL and LTE, service providers claim that the high, uneven costs for broadband services on the continent reflect the higher quality of services available. However, these claims do not seem to be true. According to the latest report of Ookla, an international broadband testing company, South Africans, for example, are on average only getting 74 % of the speeds they sign up for, which is lower than the global average of 85 %. Ookla's household promise index ranks South Africa number 55 out of 64 countries, way below Ghana, Libya, Rwanda and Kenya.

A quality of service pilot study undertaken by RIA in South Africa suggests that consumers in South Africa are not getting advertised speeds. It demonstrates that mobile outperforms fixed broadband, but that speed is not the only limiting factor on performance. Moreover, although the mobile broadband service throughput is better at all points throughout the country, it is unlikely to become a substitute for fixed-line connectivity where that is already available. Although the throughput is poor, the reliability of ADSL over time is better for those requiring consistent quality levels (Chetty et al. 2013).

Fixed Versus Wireless

So will NGB happen in Africa? We have identified some of the constraints and some of the differences between the environments in African countries and those in northern developed economies such as the United States. Why have wireless networks been so much more successful in Africa than fixed-line networks? A number of explanations have been advanced which are not necessarily mutually exclusive. The development of prepaid pricing for mobile services, greater competition in mobile than in fixed-line markets and inefficiency amongst fixed-line operators have been suggested (Wallsten 2002; Melody 2003; Gillwald 2005a, b). These reasons are not intrinsic to the technical properties of mobile technologies compared to fixed-line technologies. However, the cost structures of supplying these technologies are different. The marginal cost of adding a new user to a mobile network is in general lower than the cost of adding a new user to a fixed-line network. When considered as an investment decision—projected fixed cost investment versus projected income—this difference becomes especially marked in the case of low-consumption users. This difference applies similarly to fibre or cable networks when compared to terrestrial or even to satellite broadcast television. It is beyond the scope of this chapter to analyse this difference in detail, but it is worth noting that the most successful networks in many African countries, mobile telephony and broadcasting, including satellite broadcasting, have a low marginal cost of adding a user and that those that have not succeeded as well, electricity, fixed-line telephony and fibre to the home, do not.

This may have a direct connection with not just the relative poverty of the continent but also its physical and human geography. Development economists have pointed to the importance of geographic factors such as population density and coastal population distributions in Africa's development (Gallup et al. 1999). Telecommunication networks may be particularly sensitive to these factors. At the conference that preceded this book, Nick Rudnick, CEO of Liquid Telecom, a leading fibre network company in Southern and East Africa, estimated that while costs of rolling out infrastructure vary greatly across the continent, they are probably in the range of one-third of those in Europe, on a per-kilometre basis. But the relative cost of passing a household, a critical measure of viability for fibre networks, may be far higher in many African countries. Many African cities have much lower densities of residential households, particularly in higher income areas, than those in many European or South Korean cities. At a recent meeting of FTTH businesses in Cape Town, one industry leader stated that his costs (on a per-home-passed basis) in East Africa were similar to those in Sweden. Wananchi, the largest fibre and hybrid cable operator in Kenya, has been developing lower cost methods, importing skills and technology from the Philippines. But so far few other operators have extended these methods to other countries.

The Next Generation for Africa?

In contrast to our wireless-connected present, Noam challenges us to imagine a fixed-line (fibre-based) NGB future for Africa driven by consumer demand for rich video entertainment content. Our evidence supports his contention that demand for entertainment content may be as strong in Africa as elsewhere when measured by interest or behaviour. However it also points to two significant constraints in that demand being translated into a market capable of sustaining NGB fibre networks. First, most consumption of entertainment in the African countries reported here is free at the point of consumption, and low incomes in Africa support relatively low advertising revenues. This is a very different competitive environment for video entertainment in the United States, for example, where more than eight out of ten households subscribe to pay television services (Hollywood Reporter 2013). We have suggested that this may explain in part the relatively limited consumption of entertainment video and games online in Africa when compared to developed nations. Second, the very poor fixed-line infrastructure, both backbone and last mile, makes the investment case particularly challenging. We start from a very low base: less than one million fixed broadband connections in the whole of sub-Saharan Africa, of which more than half were in South Africa (Williams et al. 2011, pp. 262–263).

The history of network infrastructure in Africa—from the railways to mobile telephony—shows that while most if not all of these technologies have come to

Africa in some shape or form, only a few, broadcasting systems and mobile wireless networks, have become very widely available. In other cases, supply and demand have met, not at the scale of a mass consumer market but at the level of a much smaller market made up of urban elites and firms. Over a 100 years ago, the greatest network infrastructure project of the time—the Cape to Cairo Railway—did not connect many of the major urban populations of the continent, as railways did elsewhere, but rather connected the continent's major mineral resources to ports for export to European and other markets (Williams 1921). As with the railways, whether and how next-generation fibre networks get built in Africa, where and who they connect and when this happens will be determined by similarly complex interactions of economic, political and technological factors.

Our research shows that the rate at which Internet adoption in Africa is growing is dramatic, but these are off very low bases and within the constraints of the available network infrastructure and available devices. It suggests that an NGB network, that would require an entirely new last mile fibre infrastructure and new devices (including televisions), is unlikely to come to most Africans in the short to medium term. Meanwhile policymakers and market participants need to focus on what constitutes the next milestone or generation of Internet connectivity in Africa: accessible and affordable broadband. If capacity constraints to the delivery of current broadband are addressed, prices fall and investment is sufficient to address the quality of service issues, then it is possible to imagine a near future in which most Africans have access to broadband services. However, the price and quality of services that continue to inhibit the uptake of services, through whatever mode, remain highly dependent on effective regulation of the imperfect markets that characterise infrastructure industries, more so in Africa with the lack of competition in many markets. With one or two exceptions, policymakers have not yet created sufficiently competitive markets or enabled the type of open-access common carrier networks that could carry affordable (current generation) broadband services to the populace. The longer term policy implications of this are severe. The lack of development of always-on high-speed and -quality bandwidth in the access networks (last mile) required by business, public institutions and citizens has negatively impacted on Africa's informational development, a major determinant of global competitiveness.

However, even if policymakers, regulators and operators in Africa meet these current challenges, Noam's arguments remind us that what is, in global terms, "fast enough" has been and is likely to continue to be a moving target and the relationship between fixed and wireless technologies is not static. Some would argue from a policy point of view that having lower speeds widely available but with the focus on them being of sufficient quality to be used reliably for current needs may be more important than targeting speed or throughput alone. Whether and when (and over what parts of the planet) 1 Gbps speeds over fibre become "the new normal" is yet to be seen. But since its inception, Internet speeds have continued to get faster, and African Internet users have been and will be affected by these global changes. While NGB may not be "next" for most Africans, it may be for some, and on past evidence, even a couple of Internet generations may not be decades away.

Appendix: Household Survey

The RIA e-Access & Usage survey delivers nationally representative results for households and individuals using enumerator areas (EAs) of the national census sample frames as primary sampling units and sampling households from created listings for each EA. The random sampling was performed in four steps for households and five steps for individuals.

- Step 1: The national census sample frames were split into urban and rural EAs.
- Step 2: EAs were sampled for each stratum using probability proportional to size (PPS).
- Step 3: For each EA a listing was compiled, serving as sample frame for the simple random sections.
- Step 4: 24 Households were sampled using simple random samples for each selected EA.
- Step 5: From all household members 15 years or older or visitor staying the night at the house one was randomly selected based on simple random sampling.

The desired level of accuracy for the survey was set to a confidence level of 95 % and an absolute precision (relative margin of error) of 5 %. The population proportion P was set conservatively to 0.5 which yields the largest sample size (Lwanga and Lemeshow 1991). The minimum sample size was determined by the following equation (Rea and Parker 1997):

$$n = \left(\frac{Z_{\mathrm{a}}\sqrt{p(1-p)}}{C_{\mathrm{p}}}\right)^2 = \left(\frac{1.96\sqrt{0.5(1-0.5)}}{0.05}\right)^2 = 384$$

Inserting the parameters for the survey yields the minimum sample size for simple random sampling. Due to the sampling method chosen for the survey the minimum sample size has to be multiplied by the design effect variable (Lwanga and Lemeshow 1991). In the absence of empirical data from previous surveys that would have suggested a different value, the default value of two was chosen for the design effect (UNSD 2005). This yields then a minimum sample size of 768 for households and individuals. The actual sample size is slightly larger than the minimum requirement to compensate for clustering effects and to have a wide enough spread of EAs throughout a country (Table 3.9).

Weights were constructed for households and individuals. The weights are based on the inverse selection probabilities[10] and gross up the data to national level when applied.

Household weight: $\mathrm{HH_w} = \mathrm{DW}\frac{1}{P_{\mathrm{HH}} \times P_{\mathrm{EA}}}$.

Individual weight: $\mathrm{IND_w} = \mathrm{DW}\frac{1}{P_{\mathrm{HH}} \times P_{\mathrm{EA}} \times P_{\mathrm{I}}}$.

[10] See UNSD (2005) p. 119 for a detailed discussion on sampling weights.

Table 3.9 Survey summary

Target population	All households and all individuals 15 years or older
Domains	1 = national level
Tabulation groups	Urban, rural
Oversampling	Urban 60 %, rural 40 %
Clustering	Enumerator area (EA) national census
None response	Random substitution
Sample frame	Census sample from NSO
Confidence level	95 %
Design factor	2
Absolute precision	5 %
Population proportion	0.5, for maximum sample size
Minimum sample size	768
Household	Constitutes a person or a group of persons, irrespective of whether related or not, who normally live together in the same housing unit or group of housing units and have common cooking arrangements
Head of household	A head of a household is a person who economically supports or manages the household or, for reasons of age or respect, is considered as head by members of the household or declares himself as head of a household. The head of a household could be male or female
Member of a household	All persons who lived and ate with the household for at least 6 months including those who were not within the household at the time of the survey and were expected to be absent from the household for less than 6 months
	All guests and visitors who ate and stayed with the household for 6 months and more
	Housemaids, guards, babysitters, etc. who lived and ate with the household even for less than 6 months

Household selection probability: $P_{\mathrm{HH}} = \frac{n}{\mathrm{HH_{EA}}}$.

EA selection probability: $P_{\mathrm{EA}} = m\frac{\mathrm{HH_{EA}}}{\mathrm{HH_{STRATA}}}$.

Individual selection probability: $P_{\mathrm{I}} = \frac{1}{\mathrm{HH}_{m15+}}$.

DW = design weight compensation for oversampling of major urban and other urban EAs and under-sampling of rural EAs.

$\mathrm{HH_{EA}}$ = number of households in selected EA based on information of last census or updated listing by field team.

$\mathrm{HH_{STRATA}}$ = number of households in strata (major urban, other urban, rural).

HH_{m15+} = number of household members or visitors 15 years or older.

m = target number of EAs for each strata (major urban, other urban, rural).

n = target number of households in EA.

References

Adam, L. (2010). Ethiopia ICT sector performance review 2009/2010, towards evidence-based ICT policy and regulation volume two, Policy Paper 9, 2010 from www.researchICTafrica.net

Calandro, E., Deen-Swarray, Esselaar, S., Gillwald, A. M., & Stork, C. (2012). Mobile Usage at the Base of the Pyramid in South Africa, World Bank, Infodev, http://www.infodev.org/en/Publication.1193.html

Chetty, M., Calandro, E., & Feamster, N. (2013). Measuring Broadband Performance in South Africa, Policy Paper No. 2, from www.researchICTafrica.net

CISCO (2011) Cisco Visual Networking Index: Forecast and Methodology, 2012–2017. Retrieved June 17, 2013, from http://www.cisco.com/en/US/solutions/collateral/ns341/ns525/ns537/ns705/ns827/white_paper_c11-481360_ns827_Networking_Solutions_White_Paper.html.

Gallup, J. L., Sachs, J. & Mellinger, A. (1999). Geography and economic development. CID Working Paper No. 1, March 1999, Cambridge: Centre for International Development, Harvard University.

Gillwald, A. (2005a). Good intentions, poor outcomes. *Telecommunications Policy, 29*(4), 461–491.

Gillwald, A. (2005b). Stimulating investment in network development: The case of South Africa. In A. Mahan & W. Melody (Eds.), *Stimulating investment in network extension* (pp. 259–290). Montevideo: LIRNE.

Hollywood Reporter. (2013). U.S. pay-TV subscriber numbers up for 2012, fourth quarter, Hollywood Reporter, March 6, 2013. Retrieved May 28, 2013, from http://www.hollywoodreporter.com/news/cord-cutting-pay-tv-subscriber-426341.

IDC. (2012). Worldwide internet broadband bandwidth demand 2012–2015 forecast. Framingham: IDC

Lwanga, S., & Lemeshow, S. (1991). *Sample size determination in health studies—A practical manual*. Geneva: World Health Organisation.

Melody, W. H. (2003). Roadblocks on the information highway. Johannesburg: This Day, OpEd Nov 28.

Noam, E. (2008). If fiber is the medium, what is the message? Next generation content for next generation networks. Communications and Strategies, Special issue, Nov. 2008, p. 1.

Noam, E. (2011). Let them eat cellphones: Why mobile wireless is no solution for broadband. *Journal of Information Policy, 1*(2011), 470–485.

Qiang et al. (2009) Economic impacts of broadband. In C. Z. W. Qiang, C. M. Rossotto & K. Kimura (Eds.), Information and communications for development 2009: Extending reach and increasing impact (pp. 35–50). Washington DC: World Bank.

Rea, L., & Parker, R. (1997). *Designing and conducting survey research: A comprehensive guide*. San Francisco, CA: Jossey-Bass Publishers.

Research ICT Africa. (2012). Africa Prepaid Mobile Price Index 2012: South Africa, RIA Policy Brief No.1, Mar 2012, Cape Town. Retrieved Mar 17, 2013, from http://www.researchictafrica.net/publications/Country_Specific_Policy_Briefs/Africa_Prepaid_Mobile_Price_Index_2012_-_South_Africa.pdf.

SABC. (2013). SABC Television Rate Card. Retrieved May 25, 2013, from http://www.sabc.co.za/wps/wcm/connect/1e22af804f35349e9cd9fc3fdb56b4e8/Jan13+Rate+Card.pdf?MOD=AJPERES&CACHEID=1e22af804f35349e9cd9fc3fdb56b4e8.

Samsung. (2013). Samsung Announces World's First 5G mmWave Mobile Technology, Retrieved May 15, 2013, from http://global.samsungtomorrow.com/?p=24093.

Stork, C., Calandro, E., & Gillwald, A. N. (2013). Internet going mobile: Internet access and use in eleven African countries. *Info, 15*(5), 4.

United Nations Statistics Division (UNSD). (2005). *Designing household surveys samples: practical guidelines*. New York, NY: United Nations.

Wallsten, S. (2002). Does sequencing matter? Regulation and privatisation in telecommunications reforms. World Bank Policy Research Working Paper (2187)

Williams, R. (1921). The Cape to Cairo railway. *African Affairs, XX*(LXXX), 241–258. London.
Williams, M., Mayer, R., & Minges, M. (2011). *Africas ICT infrastructure—building on the mobile revolution*. Washington, DC: World Bank.
World Internet Project. (2012). The World Internet Project International Report (3rd ed.), USC Annenberg School Center for the Digital Future
Yepes, T., Pierce, J., & Foster, V. (2008). *Making sense of Africa's infrastructure endowment: A benchmarking approach*. Washington, DC: World Bank.
Yun, K., Lee, H., & Lim, S. H. (2002). *The growth of broadband Internet connections in South Korea: Contributing factors*. Asia-Pacific Research Center, Stanford Institute for International Studies.

Chapter 4
The State of Broadband in Africa: What's Here and What's Coming?

Tim Kelly

Why Broadband in Africa Matters

The first decade of the twenty-first century was a remarkable period of growth for the continent of Africa (World Bank and African Development Bank 2012). The annualized level of GDP growth between 2000 and 2010, at just over 4 % per year, was more than twice what had been achieved in either the 1990s or the 1980s (Fig. 4.1). Although there are many reasons behind this—including rising commodity prices, greater political stability, and better governance—a key factor has been the success of the information and communication technologies (ICT) sector in Africa, notably the rise of mobile communication. At the start of the decade there were just 16.5 million mobile phone subscriptions in Africa, but by the start of 2012, there were some 650 million subscriptions, or two for every three Africans, and more than in the United States or the European Union. World Bank research indicates that between 2000 and 2008, Africa's early telecom reformers enjoyed an extra 1.2 % point boost to GDP compared to those that only liberalized their telecom sectors later (Williams et al. 2011, p 111).

The direct contribution of ICTs to Africa's economy and its growth is impressive. In 2011, the mobile phone ecosystem provided more than five million jobs and contributed around US$15 billion directly to government revenues in sales, import taxes, and regulatory fees (Kearney 2011, p 21). The rising demand from consumers is also spurring greater investment in the sector. In the telecommunication sector, private investment, much of it from foreign sources, contributed some US$77 billion between 2000 and 2010 for sub-Saharan Africa. Africa is now a much easier

The views expressed in this chapter are those of the author and do not necessarily engage the World Bank or its membership.

T. Kelly (✉)
The World Bank Group, 1818 H St. NW, Washington, DC 20433, USA
e-mail: tkelly@worldbank.org

J. O'Neill et al. (eds.), *Broadband as a Video Platform*, The Economics of Information, Communication, and Entertainment, DOI 10.1007/978-3-319-03617-5_4,

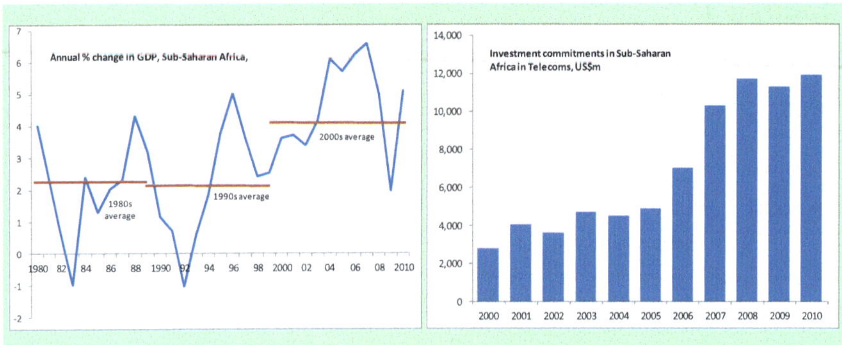

Fig. 4.1 ICT driving Africa's renaissance. Africa's economic growth, by decade, 1990–2010 *(left chart)* and private investment in telecoms, 2000–2010 *(right chart)*. *Source*: World Bank, World Development Indicators, PPI Database

place to do business, thanks to its much-improved connectivity. ICTs directly contribute up to 7 % of Africa's GDP, which is higher than the global average.

The growing popularity of mobile phones in Africa is driving the demand for bandwidth. At the start of the new millennium, the entire continent of Africa had less international Internet bandwidth than the tiny country of Luxembourg (ITU 2000). As recently as 5 years ago, the situation did not look promising, but a new generation of international cable projects has transformed the situation, at least for international connectivity, as more than a dozen submarine cable projects have connected Africa to the other rest of the world, including ACE, WACS, EASSy, and SEACOM. Some 68,000 km of submarine cables had been rolled out by 2012, and a further 92,000 km are planned.

But bringing that connectivity from the coast into the heart of Africa is a challenge that is still to be addressed. For example, take the case of South Sudan, the world's newest nation. There are no fiber-optic cables within the borders of South Sudan, and all international connectivity is provided by very small aperture terminal (VSAT) satellite dishes or by microwave. Indeed the most reliable form of communication between the ten state capitals is still the telegraph network, built a century ago. In the offices of the Ministry of Telecoms and Postal Services, in downtown Juba, the telegraph operator is the busiest man around and is happy that his skills in Morse code, learned as a freedom fighter, are still in demand (Fig. 4.2).

Even for countries with access to a coastline, Internet bandwidth can be constrained. The 800,000 citizens of Comoros have to make do, for instance, with just one 155 Mbit/s circuit. Comores Telecom, which in 2013 was still a state-owned monopoly, pays some US$2.750 per Mbit/s of international capacity, more than ten times higher than in Kenya, despite having access to the EASSy undersea cable. The monopolistic structure of the market, combined with the lack of economies of scale, makes it hard to reach critical mass in Internet use.

Fig. 4.2 The public telegraph operator at the Ministry of Telecommunications and Postal Services, Republic of South Sudan

Without fiber-based international connectivity, broadband prices in South Sudan, at over US$100 per month for a 1 Mbit/s connection, are some of the highest in Africa (Kelly and Minges 2011), but African prices generally remain well above the global average. And yet, broadband is vital to the health of modern economies. A 10 % increase in broadband is associated with a 1.4 % increase in GDP per capita in developing countries (Qiang and Rossotto 2009, World Bank, 2009). But few African countries have reached that level of penetration.

The challenge for Africa in the second decade of the twenty-first century is therefore to match, in broadband networks, the success that was achieved in narrowband mobile services in the first decade.

Where Does Africa Stand?

At the start of 2011, there were some 37 million broadband subscribers in Africa, mainly on mobile networks (Fig. 4.3). Penetration rates are high in North Africa, but raw numbers are higher in sub-Saharan Africa. At the start of 2011, approximately eight African economies had more than a million broadband subscribers, with South Africa having the most subscribers, but Libya, perhaps surprisingly, having the highest level of penetration. Prices for broadband service remain high in Africa.

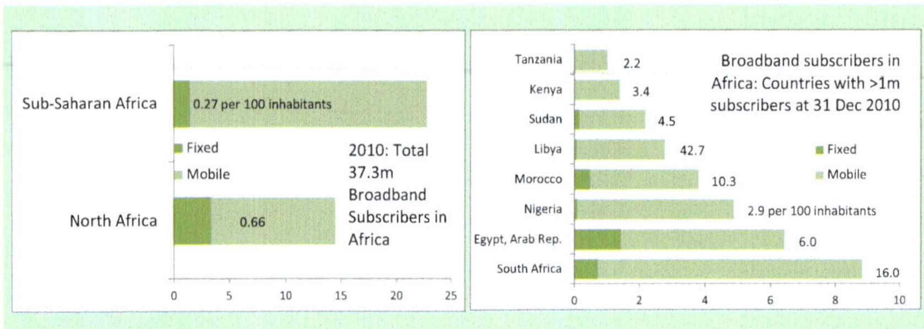

Fig. 4.3 African broadband. Broadband subscribers in Africa, and selected countries with more than one million subscribers, 2010. *Source*: ITU World Telecommunication Indicators Database

In the same way that most African's first telephone was a mobile, it is likely that their first experience of using the Internet will also be on a mobile device. Africa's adoption of broadband Internet will follow a different pattern from that of the Organization for Economic Co-operation and Development (OECD) countries, where mobile Internet came only later, after networks based on fixed-line digital subscriber line (DSL) technologies had been established. This evolution pattern has some important implications:

- Mobile broadband is likely to remain more expensive than fixed-line broadband on a per-megabit (MB) basis. Given that Africans tend to be more price sensitive than citizens of OECD countries, this suggests that usage will be lower.
- In addition, the practice of always-on, unlimited pricing, which was key to driving broadband take-up in the transition from dial-up, in the early 2000s, is much less common for mobile broadband. Although "data bundles" are common, representing a hybrid between metered pricing and data caps, mobile broadband is likely to prove expensive and unreliable for data-intensive uses, such as streaming movies or online video chat.
- This has implications for the ways in which users "learn" about broadband Internet use and information search. The best way to learn is through unrestricted use, clicking on links without much concern about price implications. By contrast, users who "learn" Internet use with one eye on the clock are likely to be more constrained in their use and hesitant to explore.
- Furthermore, most usage of mobile broadband in Africa will initially be from featurephones, with smaller screens, rather than smartphones or tablets. In consequence, Internet use is likely to be via relatively "closed" applications, rather than from "open" Internet browsers. Popular Internet uses include applications such as facebook, whatsapp, and MXit.
- Finally, because mobile devices tend to be personal rather than institutional, most first-time Internet users in Africa are likely to be paying their own way, rather than having the freedom to use the Internet at school, university, or work. This may change in time, for instance with a number of African countries

adopting laptops in schools, such as the April 2013 announcement by Kenya's new government of one laptop per child for all 7-year-olds. But in the short term, broadband use is likely to be a personal rather than a corporate expense.

Missed Opportunities in Middle-Income Countries

If broadband is to take off in Africa, and overcome these structural constraints, one would expect middle-income countries to take the lead. As might be expected, the Southern African trio of countries—Botswana, Mauritius, and South Africa—are near the top of the African league table of broadband use. But they are not performing as well as might be expected on the global stage.

The three countries represent quite distinct cases—an island state, a landlocked country, and one of Africa's largest economies. They also have similar, but nevertheless different, reasons for earmarking broadband as an important policy component in national competitiveness:

- In Botswana, the national imperative is to reduce dependence on diamonds and other mineral resources. Per capita revenues from mineral resources are already in decline since the peak they reached of around 7,000 Pula (around US$880 at 2012 exchange rates) per person. Broadband can help by improving national and enterprise competitiveness and by creating alternative and diversified sources of revenue growth in areas like financial services. Because of Botswana's landlocked position, its policies of competitiveness and development of service industries will allow it to offer lower prices and higher performance than its larger neighbor, South Africa.
- In Mauritius, broadband is part of a bigger plan to create a cyber island and to make ICT into a "fifth sector" of the national economy. In the 2013 budget speech, there is a commitment to "Embracing a technological future" and recognizing that "the Internet is a fundamental right for all our citizens" (Mauritius Ministry of Finance and Economic Development 2012). Among the measures announced to achieve this is the use of the Universal Service Fund as the tool to subsidize entry-level broadband access to bring down the price from 349 to 200 Rupees per month (i.e., US$6.35). Mauritius, in particular, wants to ensure that foreign investment is attracted to the country and that it can compete in business process offshoring (BPO) with rivals such as India or Singapore.
- In South Africa, a national broadband network is seen as a socially and geographically unifying force, with a big focus on extending the network into rural areas, even if this means the local and regional governments rather than the private sector taking the lead role. There is much admiration in South Africa for the position taken in Australia where the government has committed to underwriting a national broadband network. The plan is expected to become clearer when an updated national broadband plan is issued in 2013, superseding the current plan. The country's international broadband connectivity was seen as a critical factor in winning a share of the Square Kilometre Array radio telescope.

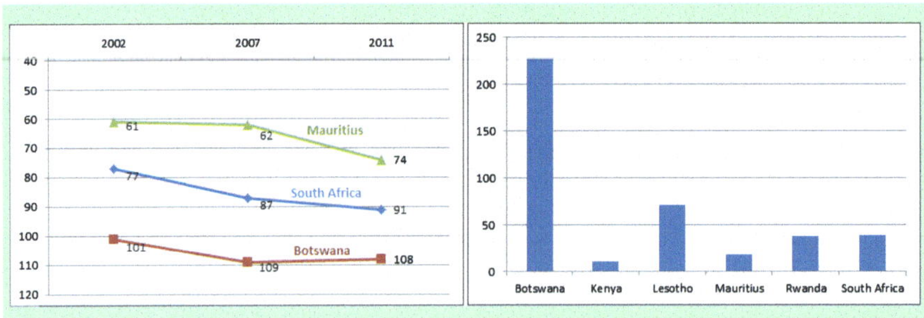

Fig. 4.4 Failure to compete. Comparative performance on the ITU ICT Development Index, and price of 1 GB of mobile broadband, Oct 2011, in US$. *Source*: ITU (2009–2011) Measuring the Information Society, and ICTdata.org

Both Mauritius and South Africa have developed national broadband plans, and Botswana is in the process of developing one. There is a strong commitment to developing broadband in each of the three countries. But, for one reason or another, success is proving elusive. All three countries are succeeding in growing their broadband networks, but they are doing so at a slower rate than their international competitors, especially in relation to the booming economies of developing nations in Asia. Indeed, Southern Africa is being outperformed even by some African nations, notably Kenya which is emerging as an ICT powerhouse with some of the lowest prices on the continent for broadband services (see Fig. 4.4).

The three economies share some key failings which have been responsible for their relative failure to develop competitive broadband offerings. Half-hearted privatization is one reason: in South Africa, Telkom was privatized relatively early (1999), but the new purchasers were then given a pronged period of exclusivity. In Mauritius, the state has still retained a controlling 60 % stake, even though France Telecom/Orange was bought in as a strategic investor. In Botswana, privatization was discussed as long ago as the early 1990s, but until now it has still not happened. The current plan is to sell a minority stake, 49 %, to local investors and the wider population, but this has been further slowed by a scheme to separate out the backbone and international cable network of the incumbent, BTC, into a state-owned entity, BOFINET.

Beyond the half-hearted privatizations, each country has also suffered from a failure to open up access to the incumbent's networks to competition through schemes such as local loop unbundling, which has proved successful in Europe. In Mauritius, which has only just over a million potential users, the competing networks of Emtel and Bharat Telecom have been obliged to construct their own backbone networks because they cannot get reasonably priced access to that of the incumbent, Mauritius Telecom. In South Africa, the fixed line network is in precipitous decline, losing more than one million subscribers between 2000 and 2011. Telkom South Africa seems to be following the strategy that if it cannot retain access to its subscribers, its competitors should not be allowed to do so either. In Botswana, the seemingly interminable delays over the creation of BOFINET, with at least four consultancy firms advising different parts of the government, have created a paralysis in the market, and the mobile companies may be obliged to construct their own networks to guarantee customer access to high-speed broadband.

A One-Time Opportunity for African Broadband: Digital Switchover

The story of African communications over the last decade is success in mobile networks, but relative failure of fixed line networks, where governments have been slow to relinquish outdated structures of state ownership and monopoly. On the horizon, though, is a one-time opportunity to benefit from the additional spectrum freed up by the transition of radio and television broadcasting to digital standards, the so-called digital switchover (DSO). Although few African countries are likely to meet the target established by the International Telecommunication Union (ITU) of 2015, there is a growing momentum for change, and there is pressure from the mobile operators to release spectrum under the so-called digital dividend (Bezzina, 2013).

The biggest impact of the DSO transition is improved television service through better picture quality, improved national coverage, and a wider choice of channels. But improvements are also expected in broadband Internet through the efficiency gains afforded by digital broadcast technology. The spectrum that the switchover liberates presents a significant opportunity for extending Internet access (through mobile broadband). The spectrum freed up by the DSO tends to be more valuable, because it uses lower frequencies that have a greater range. This is particularly important for rural areas because the capital cost for networks using these frequencies can be lower than that of existing networks which must use the higher frequencies. In theory, by 2015, a large new swathe of prime spectrum, between 694 and 862 MHz, should become available for implementing mobile broadband in Africa, and with it, the possibility of video streaming.

Of the 55 African countries, only Mauritius is currently certain to achieve the DSO early, having started the process in 2006. By the end of 2011, some 250,000 set-top boxes had been sold for the country's 350,000 inhabitants. The analogue signal is due to be switched off at the end of 2013. In anticipation of this, the government has been undertaking a consultation process and revising legislation to allow it to use market mechanisms for the allocation of spectrum, including the use of auctions and secondary trading.

Making more high-quality spectrum available promises a supply-side boost to broadband availability in Africa. This needs to be matched by innovative approaches by suppliers to convert the spectrum into affordable services, using appropriate technologies, which might be cheaper to roll out. In Mogadishu, for instance, a 1-year-old start-up, Somalia Wireless, is planning to use the so-called Super WiFi technology to reach users who currently pay an entry-level price of US $350 per month for a 256 kbit/s connection.

But supply-side measures also need to be matched by demand-side actions aimed at increasing the *absorptive capacity* of the country (Kelly and Rossotto 2012). In the context of broadband, it may be considered in terms of:

- The capacity of businesses to create broadband-enabled services and applications to improve productivity and efficiency
- The capacity of individual users to create and use broadband-enabled services and applications to improve their welfare
- The capacity of government departments and other institutions (for instance, schools and hospitals) to introduce and accommodate broadband-enabled services to deliver public services more efficiently to the public

In Africa, given the restrictions noted above on the unlimited use of broadband and the scope for experimentation, absorptive capacity is quite low. Nevertheless, there is a huge consumer demand for mobile broadband because of the lack of substitutive services that meet the need for information, such as newspapers, multichannel TV, or fixed line broadband. It is important therefore to plan for the one-time opportunity offered by the DSO with both regulatory measures, to make sure that the newly available spectrum ends in the hands of those best able to make use of it, and demand-side measures to ensure that the necessary local content and applications are available to make it interesting. The success of mobile communications has finally killed off the myth that Africa is not yet ready, or able to pay, for ICTs.

The stage is set, therefore, for a new digital revolution in Africa to follow the mobile revolution, in which narrowband networks will be converted to broadband over the course of the next decade or so. The preconditions are falling into place—improved international connectivity, spectrum liberated by the DSO, competitive markets, local content and applications development, and national broadband plans. The potential impacts are huge and transformational, especially in information-intensive parts of the economy, such as education, health, and government. Let it happen.

References

Bezzina, J. (2013). Development, the digital divide and the digital switchover: Why the DSO in Africa really matters. Retrieved from http://siteresources.worldbank.org/INFORMATIONANDCOMMUNICATIONANDTECHNOLOGIES/Resources/1400395_DigitalDivide_PolicyNote-04.pdf

ITU. (2000). *Improving IP connectivity in the least developed countries*. Retrieved from http://www.itu.int/osg/spu/ni/ipdc/study/Improving%20IP%20Connectivity%20in%20the%20Least%20Developed%20Countries1.pdf

Kearney, A. T. (2011). *Africa mobile observatory 2011*, prepared for GSM Association. Retrieved December 2011 from http://www.gsma.com/documents/download-full-report-pdf-1-31-mb/21101

Kelly, T., & Minges, M. (2011). *"South Sudan: ICT sector background", Prepared for South Sudan donor coordination conference, Washington, DC,* December 2011.

Kelly, T., & Rossotto, C. (2012). *Broadband strategies handbook*. Retrieved June 6, 2013 from https://openknowledge.worldbank.org/handle/10986/6009

Mauritius Ministry of Finance and Economic Development. (2012). *Budget speech 2013*. Retrieved from http://www.gov.mu/portal/site/MOFSite?content_id=026798395977a310VgnVCM1000000a04a8c0RCRD

Qiang, C., & Rossotto, C. (2009). Economic impacts of broadband. In World Bank (Eds.) *Information and communication for development: Expanding reach and increasing impact*. Retrieved from www.worldbank.org/ic4d

Williams, M., Mayer, R., & Minges, M. (2011). *Africa's ICT infrastructure: Building on the mobile revolution*. World Bank. Retrieved from http://siteresources.worldbank.org/INFORMATIONANDCOMMUNICATIONANDTECHNOLOGIES/Resources/AfricasICTInfrastructure_Building_on_MobileRevolution_2011.pdf

World Bank. (2009). *Information and communication for development: Expanding reach and increasing impact*. Retrieved from www.worldbank.org/ic4d

World Bank, & African Development Bank. (2012). *eTransform Africa: The transformational use of information and communication technologies in Africa*. Retrieved from www.eTransformAfrica.org

Chapter 5
The Cost of Deploying a Successful Video Broadband Business in Africa and the Cost of Not Deploying: Domestic and Transborder Implications and Applications

Ben Akoh

Introduction

Fixed broadband deployment has not experienced much growth in sub-Saharan Africa (SSA) in comparison to Europe and North America (see Fig. 5.1). Recent figures put it at approximately 2 persons in every 1,000 persons compared to 280 in North America and 260 in Europe.

Plausible explanations for this include (a) the absence and obsolescence of fixed network infrastructure in a majority of SSA countries; (b) the increase of mobile access which has taken over as the de facto communication infrastructure; and (c) the increased difficulty and spectrum requirement of using mobile technology for broadband growth and technological advancement.

Mobile broadband "leapfrogs" Africa's last mile infrastructure as indicated by recent uptake figures of nearly 4 per 100 inhabitants of active mobile broadband subscribers in Africa. However, policy, technical, and much broader developmental consideration should be paid to mobile networks in their present form if they are to carry the weight of present and projected bandwidth weight imposed by demanding applications and services.

In Williams' (2010) report "*Broadband for Africa: Developing Backbone Communication Networks*," the state of broadband deployment in Africa is conceived

The concepts in this chapter were first presented at "the Broadband as a Video Platform: Strategies for Africa" conference held in Lusaka, Zambia, on May 22–23, 2012. My appreciation goes to the Colombia Institute for Tele-Information, the International Institute for Sustainable Development, and the World Bank for funding my participation.
Ben Akoh is a Program Director at Continuing Education, Extended Education at the University of Manitoba. At the time of writing this report, he was Project Manager at the International Institute for Sustainable Development (IISD).

B. Akoh (✉)
Program Director, Continuing Education, Extended Education, University of Manitoba, Winnipeg, MB R3T 2N2, Canada
e-mail: me@benakoh.com

J. O'Neill et al. (eds.), *Broadband as a Video Platform*, The Economics of Information, Communication, and Entertainment, DOI 10.1007/978-3-319-03617-5_5,

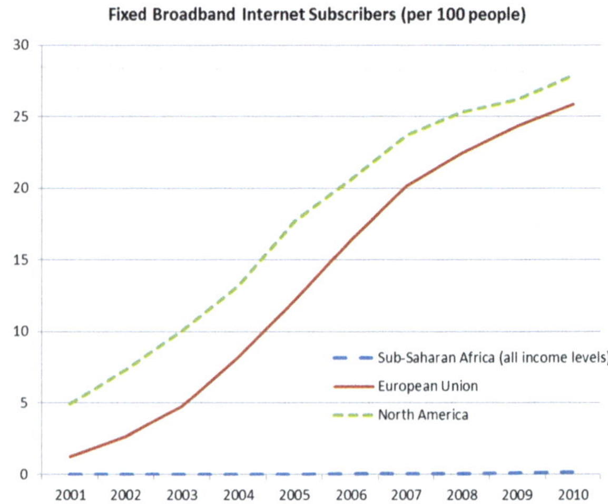

Fig. 5.1 Active mobile broadband subscribers per 100 inhabitants. *Source*: ITU (2011). Retrieved from http://www.itu.int/ITU-D/ict/statistics/material/excel/2011/Mobile_bb_11.xls

purely from a "supply and demand" viewpoint. This view argues for the "build and they will come" mindset which suggests that sufficient bandwidth supply will incentivize development and deployment of backbone infrastructure. A similar view is taken by the International Telecommunication Union (ITU 2012) and Kelly and Rissotto's (2012) leading to the argument for the establishment of (a) policies that create an enabling environment for infrastructure competition and (b) policies that stimulate rollout and deployment. As with most recent research on broadband, recommendations emanating from this view are focused on economic reforms concerning the:

- Removal of regulatory obstacles to investment and competition
- Reduction in the cost of investment
- Reduction in political and commercial risk
- Promotion of effective competition in the market
- Setup of competitive subsidy models
- Development of shared infrastructure/consortium models
- Employment of incentives based on private sector models such as universal service obligations (USOs)

This chapter argues that an economic-only point of view is necessary but insufficient to drive broadband deployment in SSA and that the cost of deployment (or not deploying) should encompass other equally important areas which may originally be perceived as uneconomically viable but are necessary even for sound economic principles to thrive. This chapter is structured into three large parts. The first part discusses the insufficiency of the present econo-centric view for broadband deployment by arguing that it places a limit on possible development of policies that are sustainable over the long run. In the second part, I present a more holistic model of viewing broadband and argue that policies should be based on other equally important development aspects than the present focus on economics.

In the final section, I conclude with three important meta-policy considerations, suggesting that during the process of defining broadband policies certain other broader considerations are required.

The Insufficiency of the Present Argument for Broadband

There are as many definitions of broadband as there are industries, interests, and opinions. One definition is Kelly and Rissotto's (2012) which describe broadband as a

> high capacity ICT platform that improves variety, utility, and value of services and applications offered by a wide range of providers, to the benefit of users, society, and **multiple** [emphasis added] sectors of the economy (p. 4).

Some, such as the World Bank take a specific econo-centric position linked purely to economic growth and development, one that assumes that development and quality of human existence, and as a consequence, life in general, are intrinsically linked to prosperity and financial growth of an economy, state, or people. This West world view may not necessarily hold true in other societies and contexts where development is considered to encompass much broader concepts than just financial.

Although broadband is an emerging area of study, most literature on the subject unfortunately also concentrates research on the economic viewpoint, to the extent that policies and positions taken by countries answer largely to the "economic development" question (Kelly and Rissotto 2012; ITU 2003, 2012; Williams 2010). Kelly and Rissotto (2012) and the International Telecommunication Union (ITU) for instance define their analysis from the impact that broadband has on Gross Domestic Product (GDP) and employment (Kelly and Rissotto 2012, p. 2; ITU 2012, p. 3).

There is no doubt that broadband deployment (video or otherwise) has made significant contributions to economic growth and development. Recent figures indicate a contribution of 1.21 % points to GDP in high-income economies and as much as 1.38 in low-income economies, more than contributions made by the Internet, fixed, and mobile telephony. Even though the data show great degrees of variance (which were explained away by methodological shortfalls, use of dummy variables, and a lack of data) (ITU 2012, p. 5), broadband has nonetheless been shown to directly, indirectly, and serendipitously contribute to job creation and economic development (ITU 2003). This can be seen in Chap. 4 authored by Tim Kelly of the World Bank in this volume.

Proponents of this econo-centric position however fracture it into two: one that favors broadband from a sub-position of volume and the other that favors it from a sub-position of functionality. These sub-positions have influenced how countries and other large international organizations base policy considerations for growth and development. Consequentially, broadband deployment as we know it today accentuates the gaps between continents, countries, and rural and urban areas.

The first sub-position, which is mostly maintained by countries and international organizations, views broadband based on speed or data transmission rate over a given period of time. As a result, a conceptual limit is placed on broadband deployment by the state. For instance, India and South Africa (DOC 2010) peg their broadband speed at a minimum of 256 kbit/s, while Nigeria's broadband policy suggests a figure of a minimum of 4 Mbits/s (BB4NG 2010) although actual figures are much lower, sometimes at 200 kbits/s. Canada, in its 2011 Broadband Report (CRTC 2011), established 5 Mbit/s downstream and 1 Mbps as target broadband to be available to all Canadians by 2015 [although it defines broadband as a minimum speed of 1.5 Mbits/s (Industry Canada 2012a)], and the US broadband goal is for affordable access of 100 Mbit/s download and 50 Mbit/s uploads to at least 100 million households in the next decade (FCC 2009). International organizations such as the ITU define broadband to mean data rates that "correspond to the user rate of 2 Mbit/s and higher." While these caps attain the objective of achievable targets, they are likewise limiting and may contribute to variable growth across countries in the same region and continent or those within the same purchasing power parity.

In a global economy that is constantly miniaturized by the growth of the Internet and in which trade, relationships, and education are required to take place between nations, capped bandwidth deployment may cripple advancements in these areas. Capping bandwidth begs answers to the questions of how a burgeoning small- and medium-scale business, say in South Africa, India, or Nigeria, with between 200 and 256 kbits/s can effectively engage with its global partner, say in Canada or the USA, with at least 50 Mbits/s. As education goes through its current phases of globalization and internationalization (concepts that describe the extension of dominant education paradigms into other society, cultures, and context) with an associated focus of educating for the global market, students with limited bandwidth would be constrained and disadvantaged from competing in this global economy. Video broadband as a next wave in content deployment requires large bandwidth. Low-capped bandwidths, especially those in SSA, would be unable to meet the demands of heavy video traffic, constituting a large negative cost to broadband deployment in Africa.

The second dominant position views broadband based on its functionality, depending on "what can and cannot be done with a certain type of connection," a position that is likewise problematic and "overly subjective" (Kelly and Rossotto 2012, p. 4). Functionality, as with the prior definitions of speed, is also restrictive and overbearing on research and development and more broadly on economic growth and long-term sustainability. Because applications and services evolve and because uses and users, not previously conceived, emerge, such positions may drive change in one direction only, resulting in large unexplored gaps in other broad areas.

Perhaps, most concerning is that an econo-centric-only view is based on a "low-hanging" fruit model which seeks the easiest path to short-term development over more rigorous longer term planning and sustainable growth. This model fractures development irrespective of the context—national or regional. In this model, incentives are provided mostly to the members of the economic sector such as the industry, which is only a subsection of the larger population. As a result, consideration for large development stops short of reaching a broader sector of society. To illustrate this point, the Government of Canada incentivizes broadband development mostly to the industry (Industry Canada 2012b). While the intent is to provide equitable and affordable access across the country, residents of large cities who can afford connectivity tend to benefit more than those in the smaller towns, reserves, and remote communities. Incentives for providing broadband access to geographically dispersed populations such as those living in the rural areas and the far north of the country are minimal compared to those living in the cities. Mignone and Henley's (2009) study of the impact of technology across indigenous Canadian communities recommends the recognition of social capital as intrinsically important for long-term sustainability of a project. Social capital is often measured in non-monetary terms and usually not considered in the deployment of technology projects, thus masking these communities from development incentives of state, industry, and investors and excluding them from innovative projects. Recognizing the social capital of local communities necessitates the design and deployment of projects that address specific needs, where those projects address longer term development concerns. Directly or serendipitously, community-based projects that recognize social capital contribute to improving literacy rates, reducing poverty, helping to achieve the millennium development goals, improving transparency in governance, and national health through for instance telemedicine projects.

If the business projections for industry are not promising, there are no incentives to extend infrastructure to these communities and, consequently, no direct or serendipitous benefits. Factors of speed and functionality linked to economic gain, which are an integral part of the current bandwidth pricing model, penalize consumers who live outside the jurisdiction of cheaper and more affordable infrastructure. While they make economic contributions to overall GDP such as in the case of a remote farmer that pays taxes but is barred from reaping the benefits of broadband infrastructure, which when considered as a general-purpose technology leaves an unfair taste behind (*see* the discussion of general-purpose technologies in the concluding section).

There is no denying that economic sustainability is a major driver for broadband deployment and development. However, this view alone could undermine broader development. Economic development, while important, is enforced by other factors such as social and environmental development. If the social factors that enhance societal cohesion are undermined, and stable environmental conditions are inexistent, elements such as civic unrest and harsh environmental and climatic impacts could, for instance, grossly affect economic growth. These other development perspectives should be taken into account and examined collectively as a whole and not in isolation as current econo-centric views of broadband have assumed. In the next section, I shall describe why such a broader and holistic view of broadband could be more appropriate.

A Holistic Model for Costing Broadband

In the previous section, I highlighted the economic context in which current broadband policies and deployment are framed. I argue that the current econo-centric view of broadband is insufficient to accomplish an all-encompassing approach to growth and development. I shall expand further on this thought in this section.

It is important to examine broadband from a more comprehensive perspective, one that transcends the economic view only. The sustainable development paradigm is useful, suggesting that for development to be sustainable, it should "meet the needs of the present without compromising the ability of future generations to meet their own needs" (Souter et al. 2010). Such paradigm transcends a low-hanging fruit model, the quick-to-deploy approach, and the capping models discussed earlier and on which most broadband development projects have been implemented and on which dominant broadband policies are based. Arguably, the per price or per function economic argument does not take into consideration these broader and longer term views because price and functions are behavioral attributes of the present. Because technology evolves ever so rapidly, an economic argument alone is a risky proposition that compromises future abilities to respond to future needs. While the question of broadband is often conceptual, in that it is difficult to define the capacity needed for a particular function in the present time or even future time, a paradigm that challenges policy makers to think beyond the present is one that will be more sustainable. The short-term cost of long-term sustainable development planning may be high, but returns over the long term is more rewarding for both the short and long terms.

In the report *ICTs, the Internet and Sustainable Development: Towards a new paradigm*, Souter et al. (2010) presented a framework first developed by the Forum for the Future as a valuable conceptual model for analyzing sustainability impacts of ICTs and specifically broadband (see Fig. 5.2). The component on the vertical axis concerns the three pillars of sustainable development, economic, social, and environmental, while those on the horizontal axis refer to the systemic effects of technology on these pillars and within any particular context.

Economic development is about "reducing and seeking to eradicate income poverty, achieving higher levels of prosperity, and enabling continued gains in economic welfare." Social development concerns "reducing and seeking to eradicate other dimensions of poverty; improving the quality of education, health, housing, and other aspects of welfare of individuals and communities; and enhancing the quality of social interaction, engagement, and empowerment." Environmental protection relates to "reducing pollution and other negative impacts on the environment, mitigating the effects of industrialization and human activity, and seeking to achieve sustainable use of resources in the interest of future generations" (p. 7).

Fig. 5.2 Sustainable development/systemic effects framework. *Source*: Souter et al.

	First order effects	Second order effects	Third order effects
Economic Sustainability			
Social Sustainability			
Environmental Sustainability			

According to the systemic effects listed on the horizontal axes, first-order effects concern "immediate and direct effect"; second order refers to "indirect effects"; and third order refers to "societal effects." Souter et al. suggests:

> The matrix can be applied to the interface between any sector or policy domain and sustainable development, but is particularly helpful to understanding the impact of ICTs on sustainable development. The table as a whole can be applied to ICTs in general, broad areas of ICT development such as the Internet or broadband networks, specific innovations such as cloud computing, or applications such as social networking. Individual cells can also be analyzed in depth, while the results of analysis can also be summarized, cell by cell, as a balance between positive and negative outcomes (p. 13).

These pillars collectively help in a better articulation of present and future requirements for broadband, more so in Africa where the environmental and social components can no longer be ignored. I will examine the cost of deploying or not of broadband using this framework focusing solely on the social and environmental pillars of the sustainable development framework given that the chapter has already argued that existing focus has remained solely on the economic aspect.

The Social Cost of Broadband (First-, Second-, and Third-Order Effects)

Typically, when corrupt government and military administrators take over traditional media in parts of the world such as West Africa, access to information and freedom of expression are muzzled. Radio and television stations are the first to be manned and used to declare *coup d'etats*. New media are shut and opposing newsprints banned from the streets. However, the proliferation of mobile phones and increasing mobile broadband have undermined these restrictive practices. Communication patterns are changing with consequential impact on how government and governance are and should be done.

In recent times, the role that technology, notably mobile phones, social media, and broadband Internet, has played in Africa's social context is consistently strained, disrupted, or strengthened, depending on what side of the government–citizen continuum one stands. From the view of government (some of which may be repressive), the proliferation of mobile Internet has increased the ability of citizens to engage in the sharing of rumors, information, knowledge, and communication between and amongst themselves, some of which are about governance and government processes. Governments are somewhat forced to provide information to citizens in order to quell impending crises or restrain the public.

In 2012 the Nigerian Federal Ministry of Information announced the use of social media as a communication tool between citizens and the government's information department during a month-long standoff after subsidy removal on petroleum products, a move which allowed government to counter citizen information. The extent to which this approach is successful requires further research; nevertheless, it highlights the importance of available broadband as the medium of communication between citizens and between government and citizens (FMI 2012).

From the view of citizens, increased interaction among themselves about government and governance demands more transparency and accountability from their leaders. Relative peace is disrupted when there is a communication gap between government and citizens. Whereas in the past, access to information was restrained and could only be attained through traditional media such as radio, television, and newsprint and only when provided by the government, citizens in Africa now are able to access information through short messaging services (SMS) or mobile Internet and are able to share video and data, some of which demand government's explanation. Freedom of expression is also enhanced, as citizens become creators of their own content and information, especially of happenings that require government intervention or happenings as a result of government intervention such as the one that triggered the so-called Arab Spring in North Africa mentioned below. This would have been less than successful without broadband Internet, particularly mobile broadband. There is a cost of deploying or not, here.

Increasing broadband has enhanced communication and communication patterns and the demand and supply of information in areas of public administration. Beyond Nigeria, the "Arab Spring" in North Africa and several other incidents in Cote d'Ivoire and Zimbabwe are indicative of the power that social media, mobile devices, and mobile broadband puts in the hands of people.

There is an opportunity here if existing governments can develop established and improved communication channels between themselves and citizens for the exchange of information and knowledge and for creating socially sustainable societies. The cost of broadband as an enabler for increased communication between government and citizens as a contribution to building and enhancing social capital that in turn enhances governance needs to be quantified. Existing bandwidth may not meet the future sustainable needs for video and voice data exchange between government and its people.

The growth of the entertainment industry is also a social driver for broadband. Local African movie production responding to an almost insatiable thirst especially by global African Diaspora is a $72 million dollar revenue-generating movie industry in Nigeria alone as in 2009 (Cartelli 2007; Madichie 2010; Shivers 2010). An industry that is third after Hollywood and Bollywood employs nearly one million people making it "the second largest employer in Nigeria after agriculture" (Barnard and Tuomi 2008). Barnard and Tuomi, in their comparative studies of the movie making industry in Nigeria and South Africa, postulate that the unsophisticated nature of movie consumers and consumption in Nigeria does not necessitate the need for production of high-quality movies. While this may be a plausible reason for increased demand in the country, it does not answer the question of increased global demand from places like the Caribbean (Cartelli 2007). These researchers argue that the distribution model for such movies, which uses low-cost DVDs at $2 a copy, necessitates increased demand over the traditional *movie-theatre-screening-then-home-video* pattern that Hollywood uses. Local consumption certainly drives demand; the suitability and ease of use of the media of distribution do contribute to this growth. However, as demand and supply increase, new methods of portable and convenient distribution may take over. Already models such as iroko.tv have placed a demand on the need for increased broadband. There is an associated cost for deployment or not, here.

Another important consideration is the demand for news and information by local audiences. While generalization may be difficult across the continent, the most popular types of television programming in the most populous nation on the continent, Nigeria, are news, musical shows, talk shows, and sports, much more than Nollywood movies (Akoh et al. 2012). The growth of other distribution channels besides DVDs such as mobile phones and Internet streaming may have contributed to this growth.

The Environmental Cost of Broadband

As it concerns environmental development, in the report *Africa Transformation-Ready: The Strategic Application of Information and Communication Technologies to Climate Change Adaptation in Africa* Akoh et al. (2012) argued the following:

1. Africa's climate is likely to be more severely affected by climate change than other regions. It is warming faster than the global average.
2. Its major economic sectors are climate sensitive.
3. Low levels of human development (income, education, health) and greater presence of other stress factors (such as conflict and disease) constrain adaptive capacity.

The report argues further that the observed and anticipated impacts of climate change in African countries continue to speak to its greater vulnerability. Addressing the impacts of climate change through the use of broadband to, for instance, develop the response capacities of citizens for sharing knowledge and information, has a cost that also needs to be quantified (Fig. 5.3).

Four areas of intervention in which ICTs can be used directly or indirectly are in the following (Akoh et al. 2012):

1. Addressing the drivers of vulnerability (e.g., the application of broadband to the effective deployment of micro-credit schemes or immunization programs): This

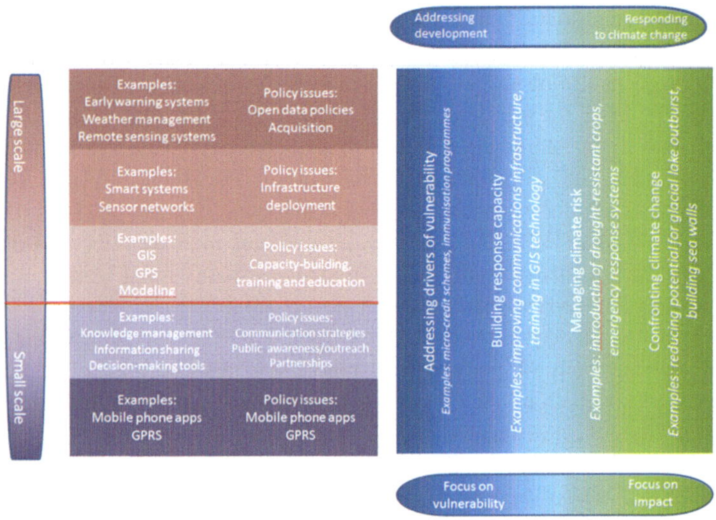

Fig. 5.3 A framework for examining ICT tools and adaptation to climate change. *Source*: Akoh et al. (2011)

intervention is development oriented, where development is directly proportional to vulnerability. Broadband deployed over mobile phones could contribute to developing the welfare of African citizens through schemes that improve their economic, health, and social well-being. These factors contribute to the choices that they can make, from which resilience emerges.

2. Building response capacity of local and regional systems and communities (e.g., using broadband to develop and deliver local-level training to community people on the monitoring of local climate variability): The availability of broadband has contributed to knowledge about climate change and the ability to manage data, information, and local and regional knowledge bases that can be used to enhance the response capacity of citizens. Whereas long-term weather and climatic information may not be readily available, and where it is in limited quantities, citizens are able to augment these with locally generated and stored information and knowledge. Availability of extensive broadband could contribute to enhancing such capacity.
3. Reducing and managing risks related to climate variability and climate change (e.g., using broadband in emergency response systems): Early warning systems generated through satellite data or sensor networks located in different locations that feed critical information to citizens on mobile devices could enhance their abilities to manage and reduce the risk of climate change. Sufficient broadband may be needed to send images and sound that enhances the quality of these decisions.
4. Confronting climate change (e.g., using broadband to manage potential flood outbursts): Data generated from buoys or sensors strategically located to warn of imminent climate-induced happenings could be sent in order to elicit the

appropriate response before they happen. For instance, sensors that send data about the amount of upstream impurities could elicit a response before it gets downstream, or systems that warn of flood could result in the shoring up of the coastal line prior to it happening.

ICTs that can be used range from small-scale applications such as mobile phone apps or online knowledge management and information sharing tools to large-scale deployment such as early warning systems, weather management applications, remote sensing systems, and sensor networks. These applications, if implemented now and with a view of future sustainability, could drive the demand for greater bandwidth through, for instance, large-scale information and data demand by industry needed to make investment decisions in any geographic location or by which citizens decide what crop to plant where. Inherently lies an opportunity for future growth and development, if the continent is to address its present and future vulnerable environmental situations. Broadband development and deployment costs need to be quantified here too.

Transborder Considerations

When social and environmental concerns such as those that have been mentioned are taken into development considerations, the costs of deploying bandwidth ought not to be borne by just one country, but by several countries in a region or across the continent. Climatic variations are not sharply contrasting across borderlines. The key drivers for deployment which include local backbone infrastructure, trans-boundary implementations, and so on are also applicable across countries as much as they are at the national level.

One significant indicator is the historic trend of negotiating for bandwidth across the continent. The shift from a single-cable to the current multi-cable system in Africa in less than one decade, and from which Africa benefits today, offers a lesson on how to address long-term transborder infrastructural development. Multipurpose vehicles between different groups of stakeholders across borders facilitated the move to the present structure, thereby strengthening the argument for collaborative transborder broadband agreements. Undersea and terrestrial infrastructure is one out of the many others that are needed. Satellite, early warning, and sensor networks are larger scale systems that could be shared across borders. Consortiums of countries and their national telecommunication companies bore the cost and risks associated with submarine cable deployment; this model could also work for these larger scale systems.

Transborder considerations such as right of passage to coastal gateways especially for landlocked countries or those that may have not invested in undersea cable infrastructure but for whom they hold economic and social promise are similarly important. The volatile nature of Africa's boundaries especially during crisis suggests that one country should not bear the cost of running and maintaining its own infrastructure and that opportunities for shared resource could support longer and sustainable peace in the region.

This is even more true given very porous boundaries between most countries in the different regions in Africa. Already the mechanisms for regional cooperation exist such as ECOWAS in West Africa, COMESA in East Africa, and SADC in the Southern African region, which facilitate transborder trade and migration but could be useful also for advancing the dialogue on collaborative regional responses to climate change or social cohesion. The cost may include sharing a portion of regional bandwidth, human resources such as experts, data, information and knowledge bases, and sometimes response capacities and vulnerabilities. The justification for shared resources has to be made by the different countries in the region. The cost of deploying broadband could form the basis for such dialogues. In the case of Spring Brook in Australia for instance, over 200 individual sensors were installed to monitor temperature and rainfall, an additional 125 sensors were installed to monitor carbon dioxide and cloud cover, and a further 175 nodes were installed to monitor bioacoustic sounds and videos; the collective data generated from these devices provides a richness of information that helps to preserve the environment and further helps in monitoring the restoration of the environment and returning wildlife (Queensland Government 2012). The data from these sensor nodes will require underlying broadband technology for its distribution and analysis. The outcome of this type of application bears on the economic, social, and environmental health of society.

Considering Broadband as a General-Purpose Technology and Its Associated Cost

One major policy argument for broadband deployment in the economic setting is for a consideration as a general-purpose technology (Kelly and Rosotto 2012; Souter et al. 2010). GPTs are technologies that enable new and different opportunities across an entire economy, rather than simply addressing one problem or one sector. It is technology that "initially has much scope for improvement and eventually comes to be widely used, to have many uses, and to have many … technological complementarities" (Crafts 2004, quoting Lipsey et al.). The steam engine, ICTs in general, and electricity have been considered as GPTs (Crafts 2004). Recent arguments have elevated broadband to the same level as GPTs because they exhibit three characteristics, namely, the following (Souter et al. 2010):

1. They are pervasive in their use in a wide range of sectors (including sectors that affect social and environmental development).
2. They are technologically dynamic (i.e., they have an inherent potential for technical improvements).
3. They provide general productivity gains as GPTs evolve, improve, and spread through the economy.

The consideration for broadband as a GPT puts it at a level where government and all stakeholders will be obligated to consider its development as important as other GPTs. Because the impact of broadband transcends the economic perspective, any consideration as a GPT should expand its scope beyond this aspect also.

Conclusion

Broadband, particularly on the mobile platform, has been on the increase in SSA with consequences on economic growth and development. Broadband policies have been focused on econo-centric aspects such as the removal of investment barriers and the promotion of competition in the marketplace. Little research and consideration have been placed on development that could occur in the social and environmental aspects of society and how these in turn affect the economic and broader well-being of a nation, region, or context. Development that considers a holistic view of broadband policy and deployment recognizes the social capital inherent in the development context and thus directly or serendipitously produces results that are more long term and sustainable. More research and specific cost implications are required in this area.

Proponents of broadband policies in their current form should step back and conduct a meta-analysis by responding to how this piece of technology directly, indirectly, or serendipitously affects development that is not merely economical but also social and environmental. The current view of elaborating broadband cost is limited and thus does not provide a holistic view of total deployment cost required. The cost of deploying broadband surpasses the economic view only. It should be considered alongside the social and economic dimensions of sustainable development. As a result, the present focus of pegging broadband by speed and functionality which is mostly linked to economic sustainability dimensions is largely insufficient to meet the more holistic demands resulting from these other components of sustainable development. A useful framework is presented that can guide policy makers in formulating broader and more sustainable policies; in incentivizing development investment for the private sector that creates employment and that recognizes the social capital and environmental health of a context; and in contributing to present and future development.

The argument for broadband as a general-purpose technology makes it relevant to broad development policies. Finally, given the broad scope of environmental concerns and the increasingly energetic and highly resourceful societies in SSA, the costs of deploying bandwidth ought not to be borne by just one country, but by several countries in a region or across the continent.

References

Akoh, B., Bizikova, L., Parry, J., Creech, H., Karami, J., Echeverria, D., Hammill, A., & Gass, P. (2011). Africa transformation-ready: The strategic application of information and communication technologies to climate change adaptation. In *World Bank*. Winnipeg: IISD. Available from http://siteresources.worldbank.org/EXTINFORMATIONANDCOMMUNICATIONANDTECHNOLOGIES/Resources/282822-1346223280837/ClimateChange_Fullreport.pdf

Akoh, B., Jagun, A., Odufuwa, F., & Akanni, T. (2012). *Mapping digital media: Nigeria*. Open Society Foundation. Retrieved from http://www.opensocietyfoundations.org/sites/default/files/mapping-digial-media-nigeria-20120813.pdf

Barnard, H., & Tuomi, K. (2008). How demand sophistication (De-) limits economic upgrading: Comparing the film industries of South Africa and Nigeria (Nollywood). *Industry and Innovation, 15*(6), 647–668.

Broadband for Nigeria [BB4NG]. (2010). *The framework*. Available from http://bb4ng.forum.org.ng/node/13

Canadian Radio-television, & Telecommunications Commission [CRTC]. (2011). *Broadband report*. Available from http://www.crtc.gc.ca/eng/publications/reports/broadband/bbreport1111.htm#n1.0

Cartelli, P. (2007). Nollywood comes to the Caribbean. *Film International, 5*(4), 112–114.

Crafts, N. (2004). Steam as a general purpose technology: A growth accounting perspective. *The Economic Journal, 114*(295), 338–351.

Department of Communications [DOC]. (2010). *Policies and policy directives drafted in terms of section 3(1) of the electronic communications act, 2005* (Act no. 36 of 2005). Available from http://www.info.gov.za/view/DownloadFileAction?id=127922

Federal Communications Commission [FCC]. (2009). *National broadband plan: Connecting America*. Available from http://www.broadband.gov/plan/executive-summary/

Federal Ministry of Information, Nigeria [FMI]. (2012). Tweet meeting. *Federal ministry of information Nigeria: Facebook page*. Retrieved May 21, 2013 from https://www.facebook.com/FMINigeria/posts/292416917477890

Industry Canada. (2012a). *Broadband Canada: Connecting rural Canadians*. Available from http://www.ic.gc.ca/eic/site/719.nsf/eng/h_00001.html

Industry Canada. (2012b). *Broadband Canada: Connecting rural Canadians—List of projects by province*. Available from http://www.ic.gc.ca/eic/site/719.nsf/eng/00050.html

International Telecommunication Union [ITU]. (2003). *Promoting broadband: Background paper*. Available from http://www.itu.int/osg/spu/ni/promotebroadband/PB03-PromotingBroadband.pdf

ITU. (2012). *Impact of broadband on the economy*. Retrieved from http://www.itu.int/ITU-D/treg/broadband/ITU-BB-Reports_Impact-of-Broadband-on-the-Economy.pdf

Kelly, T., & Rossotto, C. M. (2012). *Broadband strategies handbook*. DC: Worldbank. Retrieved from https://openknowledge.worldbank.org/handle/10986/6009

Madichie, N. O. (2010). The Nigerian movie industry "Nollywood"—A nearly perfect marketing case study. *Marketing Intelligence & Planning, 28*(5), 625–649.

Mignone, J., & Henley, H. (2009). Impact of information and communication technology on social capital in aboriginal communities in Canada. *Journal of Information, Information Technology, and Organizations, 4*, 127–145. Retrieved from http://130.179.16.50/faculties/human_ecology/departments/fss/media/5.pdf

Queensland Government. (2012). *Wireless sensor networks: Springbrook*. Retrieved March 29, 2013 from http://www.qld.gov.au/environment/plants-animals/herbarium/sensor-springbrook/

Shivers, K. N. (2010). Negotiating identity in transnational spaces: Consumption of Nollywood films in the African diaspora of the United States. In *MeCCSA Conference, London*.

Souter, D., MacLean, D., Akoh, B., & Creech, H. (2010). *ICTs, the Internet and sustainable development: Towards a new paradigm*. Winnipeg: IISD. Retrieved from http://www.iisd.org/pdf/2010/icts_internet_sd_new_paradigm.pdf

Williams, M. D. J. (2010). *Broadband for Africa: Developing backbone communications networks*. WB. Retrieved from http://www.infodev.org/en/Publication.526.html

Chapter 6
Implications of Broadband Video on National and International Media Systems, Including Growth of Local Content

Olivier Nana Nzepa

Introduction

> ... 99.9 per cent of what Europeans know and publish about Africa is false, because they are being fed with the wrong information about the great continent that remains unique in its culture, tradition and customs (Anikulapo-Kuti African political activist musician during his classical performance in Osho S.A., 2010).

> Video will be the next big thing of the Internet Industry predictions say that by 2015 more than 90 percent of web traffic will be video (Christian Kaufmann).

> The demand for highly-valued and relevant content is a major driver for Internet take-up. But the primary sources of content in Africa remain international and there is a paucity of local content available.

> Africa's focus, thus far, on mobile networks to address an immediate service need has left backbone networks underdeveloped. This has created a major bottleneck in the rollout of high-bandwidth services and in the upgrading of cellular networks to provide value-added services. Overcoming this infrastructure hurdle is an important element in shaping the structure and policy framework of the telecommunications services sector (Moshen A. Khalil. Broadband for Africa. Developing Backbone Communications Networks, The World Bank Group, Infodev, 2010). Mohsen Khalil, Director of Global ICT at the World Bank.

The hurdle is not only infrastructural. It is also political and regulatory. In a book to be published titled "*Access and Usage: is mobile telephony another plight for Africa?*" we share the view that absence of political direction, inefficiency of regulatory boards, lack of powerful consumer associations, and technology fetishism have transformed mobile service into an accelerator of poverty for more than

O. Nana Nzepa (✉)
Advanced School of Mass Communication, University of Yaoundé II, Ngoa Ekele, Yaoundé, Cameroon
e-mail: nzeps@hotmail.com

J. O'Neill et al. (eds.), *Broadband as a Video Platform*, The Economics of Information, Communication, and Entertainment, DOI 10.1007/978-3-319-03617-5_6,

50 % of African users. We therefore caution the decision makers to approach the technology deployment with less fetish-based beliefs and consider the externalities and local impacts.

The term "media system" is frequently used in ways not clear enough to provide an unambiguous definition. From media studies, we approach the media system as a complex and autonomous entity, part of a greater set, being a continent or a country, considered as a system. Such media system is therefore composed of sub-systems such as institutional structures, informal arrangements, and end products used directly or frequently by a variety of recipients. In Africa traditional media and the use of modern and Internet-related media are crucial to the functioning of the media system. That is, taking into account the uniqueness of forms of content is essential.

The local content as envisaged here goes beyond standard media, to include websites in local languages and broader local contexts. The concept encapsulates local cultures, software, government services, and educational materials hosted locally within the African continent's local loop network.

The deployment of the first generation of ICT networks has led to what the economist termed "the mobile marvels in Africa."[1] The "mobile marvels"[2] transformed the perception of Africa. While long viewed as a marginal market, the continent is becoming the new business frontier (Fig. 6.1).

Gearing towards broadband networks is a prerequisite to fulfill the unmatched needs for more, new, and affordable services. Accenture (2011) states that "fast-developing convergence of the Internet and TV ... is revolutionizing the broadcasting industry and has the potential to transform the market for every player"[3]—both established and those that are just emerging in this exciting and dynamic environment. For a continent where the broadband access is less than 1 %[4] this represents a paradigm shift, for which the recipes that led to the surge of mobile telephony might be inapplicable.

This chapter is an attempt to take stock of the broadband video revolution so as to assess its implication on national and international media systems, including growth of local content in Africa. Huge investments are being made in optic fiber deployment all over Africa. Without enough relevant local content, international content alone will not maintain usage at levels that are economically sustainable.

It also discusses the conditions under which economic models could be implemented to sustain content-thirsty broadband deployment in Africa. In addition, the chapter seeks to underscore the necessary drivers needed to frame the policy, technology, and business trends that will reshape the broadband deployment for video purposes in Africa. The assumption put forward is that upgrading existing

[1] The Economist, Mobile marvels, Sept. 24, 2009.

[2] The Economist, 2009.

[3] www.accenture.com/.../insight-bringing-tv-life-race-dominate.

[4] ITU The world in 2010: ICT facts and figures.

Fig. 6.1 *The most promising mobile markets* In: The Economist: The mobile marvels, 2009

systems to the level of next-generation, video-capable broadband is a condition critical to sustain the information society in Africa.

Lessons learned from the surge of mobile telephony and research carried out on policy formulation, access, and usage patterns enable us to anticipate that Africa is on the path to broadband deployment and adoption. Africa will face a number of challenges: infrastructural, technological, regulatory and policy formulation, and related economic models.

Broadband Video Within the African ICT Context

Africa is entering the race for broadband with a track record of successful mobile telephony technology implementation. In most African countries, the mobile networks have filled the gap created by the fixed line network sub-capacity. The success of mobile is reflected in the number of mobile phone subscribers which has grown from 4.19 per 100 inhabitants to 41 on a continental average as of 2010, according to ITU. But with more than 500 million subscribers, we have been assessing that the mobile voice market is reaching its saturation level.

Thus, mobile operators have turned towards data services as a means of capturing new sales. The problem is that in Cameroon, for example, the fixed (wired) broadband connection costs an average of 1,990 PPPs per month, as compared to barely 28 PPPs per month in developed countries. Cameroonian broadband users pay 3,000 times more for service than in Monaco. Put another way, the cost of access represents 0.3 % of average income in Monaco and 1,000 % of average income a country like Cameroon and similarly situated countries in sub-Saharan

Africa. Indeed, 11 out of the 15 so-called least developed countries economically are located in sub-Saharan region in Africa.[5]

A user in the USA pays an average of $3.33 per Mbps and a user in Japan pays $0.27 per Mbps, while a user in Nigeria pays $2,400 and a user in Kenya pays $700 for the same capacity. This imbalance explains why the mobile sector has created wealth for mobile system owners without development for countries. Income generated has been insufficient to generate employment, and the jobs that are generated are mostly in the informal sector. The cost of access remains among the highest in the world. A 2012 household and individual survey on ICT access and usage in 12 African countries demonstrated that most of the users spend well above 12 % of their income on nonrevenue-generating telecommunications services, spiking for some Africans at more than 50 % of their monthly income.[6] This data demonstrates the extent to which an environment that is not conducive to secure investment and fair competition in a transparent regulatory structure has created in some countries failed policies and broader disparities between rich and poor—creating rather than fighting against poverty (Fig. 6.2).

However, the use of mobile infrastructure to provide Internet services is encountering a number of difficulties. The growing need for more services and larger bandwidth is outstripping the existing capacity, and investment is not following fast enough to meet necessity.

For *Hamadoun Touré,*[7]

> Broadband is the next tipping point, the next truly transformational technology. It can generate jobs, drive growth and productivity, and underpin long-term economic competitiveness. It is also the most powerful tool we have at our disposal in our race to meet the Millennium Development Goals, which are now just 5 years away.

Broadband definition varies. Traditionally, it is defined as an Internet connection speed of 256 kilobytes per second (kbps). It can be provided by three types of connectivity: fixed wire line, fixed wireless, and wireless. It can be delivered by a range of access technologies: DSL, WiMAX, CDMA, EV-DO, HSPA, and LTE. For an increasing number of researchers, the speed is not an issue. With advances in technologies, the speed is always changing. They suggest instead considering broadband in terms of an "always-on" connection where the user does not need to dial a phone number to go online. "Broadband also means high-capacity networks that can deliver very large amounts of information simultaneously. As a result, they can deliver voice, data, and video, all at the same time" (quoted in Akpan-Obong and Alozie 2011:6). Most definitions of broadband transcend bandwidth and speed. They capture the "entire 'eco-system' of Internet and data services from both a demand- and supply-side perspective." Their definition is close to the World Bank's

[5] ITU StatShot—Issue 8, January 2012.

[6] Towards an African e-Index: Household and Individual ICT access and Usage across 17 African countries, edited by Alison Gilwald, 2010.

[7] *Hamadoun Touré, Secretary General, International Telecommunication Union*, 2010, quoted in Akpan-Obong and Al, 2011.

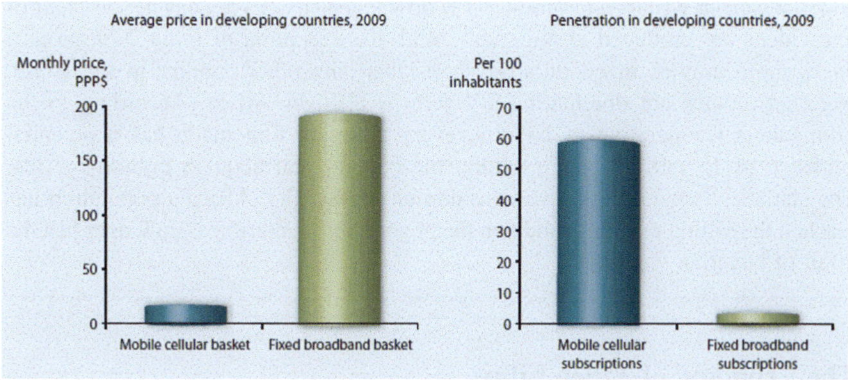

Fig. 6.2 Comparison of average price. Source: Measuring the Information Society Report 2010, ITU. Nevertheless, as mobile networks have by far the greatest per capita penetration in Africa, and as broadband is essential to Africa's struggle to continue to bridge development gaps that isolate Africa in the world economy, logic dictates that mobile infrastructure will be the workhorse to drive next-generation services in Africa

definition of broadband as "an interconnected, multilayered ecosystem of high-capacity communication networks, services, and applications." As an ICT ecosystem, broadband has three components: infrastructure, applications and services, and access.

Dynamics of Video Platform Networks

For video and entertainment purposes, the existing backbone networks are extensive in reach but predominantly low in capacity. They have been designed to carry voice communication services. As a result, the current network infrastructure is not able to carry the volume of traffic that would be generated if affordable broadband connectivity were available on a mass-market basis. And they cannot effectively carry video.

The deployment of the envisaged backbone networks with video-enabled platforms will set the stage for a rapid uptake of digital video services and fuel the growth. IP video delivery will provide a big market opportunity for a range of companies in the broadcasting, communications, cable, high-tech retail, entertainment, and content development industries.

The growing size of Africa's population and the diversity of its cultures combined with the raw talents that abound on the continent create the opportunity for phenomenal growth in the video and entertainment industries. The Nigerian film industry is a resounding example of what Africa could achieve with the right tools. In an issue of Africa Update, Gloria Emeagwali states, "The prolific output of

Nigerian screen writers and producers is now legendary. Every day at least four or five videos are produced in Nigeria."[8] With the exception of India, Nigeria produces more movies in quantitative terms than any other country in the world. Nigerian movies are dominating TV screens all over Africa. According to the Filmmakers Cooperative of Nigeria, every Nigerian film made has a potential audience of 15 million people within the country and about 5 million outside. The statistics look conservative considering the fact that Africa's population has reached the billion threshold and that the Nigerian film industry is no longer just the affair of Nigerian viewers.

The Economy of Video Films

To appreciate how this affects Africa's surge to future economic growth, consider the following. The State of California has a gross domestic product of $1.4 trillion and is the fifth largest economy in the world, richer than the combined wealth of all the 54 African countries. The wealth of California comes from the Hollywood film industry. Hollywood has generated a commercial model through which a film's commercial life-span normally begins with a box office or cinema release, then video release, then broadcast on fee-paying television, and finally on public television. The appropriate promotion and publicity enable producers and marketers to maximize profitability out of each phase. This goes for the content as well as for the supporting sound tracks in many cases. The model has made the American movie industry the second largest export revenue earner, after the aviation industry.

Another country has mastered the model: South Africa. In South Africa, according to entertainment experts, the video distribution usually doubles or triples a movie's revenues. The video boom is becoming an African phenomenon. The FESPACO, a Pan-African Film and Television Festival of Ouagadougou (which is held in the capital of Burkina Faso every 2 years) is helping a great deal to improving the quality and global appeal of Francophone films and contributing to their respective economies. In Africa, video has imposed itself as the home entertainment mainstay. The video industry is creating jobs, generating income, and contributing to the improvement of African image. Through video movies, Africans are experiencing a cultural connect worldwide.

The South Africa's leading satellite TV company, Multichoice DSTV, has caught the waves by introducing its Africa Magic channel which attracts over 1.5 million subscribers in Africa, Europe, and the Middle East. Broadband needs the video industry just as much as the video industry needs broadband networks. The African video industry is viable and has all the elements of being sustainable over the long term. And all of this would double and triple if the regulatory bodies would

[8] Africa Update, www.web.ccsu.edu/afstudy/upd11-2.html, consulted on May 12, 2012.

organize more transparently and intelligently to force competitors to become market dominators, like Multichoice.

The potential has drawn into the market an increasing number of new players, striving to explore ways and means of improving high-speed connectivity in Africa. There are significant projects currently under way in several countries in the region. Some telecommunication companies are linking up their networks between countries to establish regional backbone networks. New carrier operators are emerging, specifically aiming at providing high-capacity backbone services to other companies.

From Talking Drum to the Internet: African Media Systems at a Crossroad

As noted at the outset, the African media system is one of the most complex ones in the world. Prior to western colonization, Africa's traditional media system comprised callers, criers, and bards who served the purpose of message conveyance. The messengers were in charge of getting people's attention through various means or tools. The marketplaces served as important settings for communication.

The colonialism era brought the introduction of modern technologies that have produced today's media systems. The newspaper, the radio, and lastly the TV were introduced and used more as a political tool than as a tool for information dissemination or entertainment. Those technologies did little to vitalize the traditional media system. They accentuated the divide between the towns (homes of the colonizers/settlers) and the rural areas with villages (where the majority of the natives lived).

The African media have undergone tremendous changes in the last decade. The shift towards democratization has led to media diversity in a way never seen before in Africa. According to the UN, alternative and new forms of media, being community or privately owned commercial media, have steadily emerged and grown in numbers and diversity of opinion.[9] Consequently, while the new media entities were emerging, state-owned media, faced by competition and diversity, has stagnated.

African scholars are reflecting on the impact of digital and convergence technologies on the media systems. Questions are raised about conventional regulatory telecommunication frameworks. Research on the telecommunication regulatory environment (TRE) carried out by the Research ICT Africa Network (RIA)[10] has persistently demonstrated that the regulation in Africa to deal with new and emerging technologies lags behind.

[9] UN report, 2006 quoted by.

[10] ResearchICTAfrica.net.

The prospect, hope, and expectation are that broadband will free ordinary citizens from the hurdles of access to frequencies and licenses by making affordable digital and video cameras and audio recorders and players available to ordinary citizens. More people will be able to create, publish, podcast, stream, and share their content on the Internet. The broadband revolution will also challenge the traditional definition of journalism. The media systems as they exist today are adapting. Citizen journalism demonstrated its transformational power during the Arab Spring.

Among the number of challenges that stand in the way of a strong media sector, sustainability emerges as an overarching concern. The financial sustainability will remain a dream as long as the media system market is thin, the buying power low, and the advertiser base weak. In most African countries, there is an overall deficit of investment. Owners and entrepreneurs have little or no dedicated means of support. Most rely on donors to provide funds and support training. Such piecemeal, small-scale support cannot lead to the creation of a vivid, self-supported, and financially sustained media sector. There is a need for an economic model to assure consistent and coherent financing. Only the development of and commitment to support a sustainable economic model will free the African media systems from reliance on donor funding and lead to full sustainability.

The Ethiopian Airlines is an example of a model to consider. Ethiopian Airlines has created markets outside of its national base. The overwhelmingly urban concentration of the existing media system does not serve the national developmental purpose. The broadband revolution is bringing in its wake hope of transnational media markets, with improved capacity and standards and high-quality, accurate, reliable, pertinent, and content production. The advent and the surge of ICT and especially the Internet have changed the African media system landscape. Progress has been made. Nevertheless, the African mass media remains plagued by acute problems including a lack of financial, human, and material resources and a dearth of local content.

Local Content: A Driver

Despite the efforts related above, nearly all Internet content is hosted outside Africa. The situation has resulted in a dependency on international backhaul to access this content. There is no need for a user based in California or New York to cross the US shores or traverse its states to get cnn.com. CNN hosts its content in Atlanta. For Atlanta user, CNN content is accessible by a local connection, and in other states, it is accessible by any number of technologies that drive connection. Its user doesn't need international capacity. The users of the area will do few hops to access hosting servers for most of the content they seek. Their Internet experience is completely different from what an African user experiences. A Kenyan user has to leave the continent to access nation.co.ke hosted at Verio.net on their servers.

Internet access is therefore more expensive for the African user because he or she always has to traverse international links that are private commercial ventures.

In order to cut down cost, two conditions are critical: deploy broadband capacity networks and roll out local content hosted within Africa. Our local content concept is much more than media. Media content needs to encompass African cultures reflected in the various talking drums, folk songs, drama, festivals, town criers, traditional wears, artifacts, artworks, paintings, stories, cultural architecture that reflects African vibrancy in the palaces, shrines, and African cities, towns, and villages. What is needed are software, hardware, government services, and educational materials hosted by reliable data centers within Africa. With more than one billion people of diverse culture, tribe, ethnic, language, tradition, and customs, the uniqueness of Africa's local content is embedded in the originality, creativity, tradition, and culture of its peoples. The mobile operators are trying to cash in on this phenomenal economic potential. With voice service adoption reaching saturation, mobile service companies are seeking to boost their usage numbers by developing content for specific African regions. Various strategies are being tried to stimulate the production of more local content for their customers.

Among the most popular approaches is the establishment of mobile application stores. In an attempt to promote locally relevant mobile applications created by African developers in East Africa, Vodacom, Safaricom, and MTN have launched stores. They are also providing greater support to local application developers. Safaricom and Vodafone launched the Betavine Platform in 2011. The purpose of the platform is to help local developers design and test their applications. Safaricom has also established an academy in partnership with Vodafone and Strathmore University. The aim is to offer tailored Masters-level training in mobile software application development specifically for the African market.

MTN Play seeks to serve another purpose: promote local content through a portal showcasing an array of downloadable local content, videos, and music. The South Africa-based operator has signed a deal with the music channel Trace to enable its younger customers access tailored local multimedia and news. Safaricom similarly offers content particularly aimed at its Kenyan audience through its web portal Safaricom Live.

These various efforts seek to curb the predominantly international content available through news and web portals, such as CNN, BBC, or Yahoo, and by global Internet platforms, such as Facebook and Google's Gmail. The use of mobile infrastructure to achieve this purpose has drained the resources of the existing networks. With their capacity to sustain the high demand, cut user costs, and lay the ground for sound local content development, land-based broadband video networks, providing the backbone to mobile broadband, may be able to provide a better and more sustainable solution.

Conclusion

On its paths to broadband revolution, Africa needs to overcome three major challenges:

1. The "build-it-and-they-will-come" theory: The continent is deploying hundreds of thousands of kilometers of fiber-optic cables on pure speculation. There is a lack of strategic market research or needs assessment information. The risk is to drain the means useful for development projects into networks which could become "cyber elephants." If the expectations are not realized, the companies may not realize a return on their investment and the institutions in charge of such broadband management may not achieve the desired growth. Or this nascent industry bubble might burst, like in the US dotcom bubble or the well-known casualties of the past like WorldCom, Tyco, Global Crossing, Adelphia Communications, and of course, the railroads of yore.
2. A fragmented market: The fragmented nature of the African market contributes to the difficulty in building a useful economic model. Vinoth Gunasekaran and Dr. Fotios C. Harmantzis from Stevens Institute of Technology Telecommunications Management (2005)[11] say that there are numerous broadband business models proposed. But it is necessary to figure the most suitable to the context. Restarick and D'Aleo (2009) quoted the computing pioneer, Alan Kay, as saying that "... after the printing press: the development of a ubiquitous, high-speed broadband network [will link] people, businesses, educational institutions, services, everyday objects and appliances, and much more. Consumer demand for high-speed networks is strong, driven especially by mobile broadband connectivity. But effectively monetizing broadband deployment will be a challenge. Competition will only intensify, especially between traditional providers and cable companies."[12]

Given the challenges, it is imperative that states and investing companies start with a careful network plan and strategize accordingly. For the planners, several considerations should be taken into account:

- *Local capacity and demand for the access network.* For both broadband wireless access and fiber deployments, linking the rollout plan to local demand is the key to establishing an adequate business case.
- *Backhaul and core capacity.* Increasing the footprint and capacity of access networks requires detailed analysis and preparation of aggregation points in the network, specifically backhaul from regional exchanges and, generally, multiprotocol core networks.

---

[11] Migration to 4g-.ubiquitous broadband.economic modeling of wi-fi with wimax. Stevens Institute of Technology Telecommunications Management (Castle Point on Hudson) 2005.

[12] Planning an effective high-speed broadband deployment, Jonathan C. Restarick and Marco D'Aleo, 2009.

- *Costs for a next-generation operations support system (OSS).* The OSS structure for a next-generation network is radically different from a traditional network. Provisioning requirements and new ways of managing customers and services necessitate an enhanced OSS capability if multibillion-dollar broadband investments are to deliver adequate business value.
- *Government subsidies.* Governments see high-speed broadband networks as a key to achieving important policy objectives, so they have to take a variety of supportive actions. These can be, as in the case of Japan, the use of tax incentives and DSL regulation to drive growth in fiber-to-the-home deployment. In general, the business case for many carriers will depend heavily on direct or indirect government subsidy and on the direct participation of local municipalities.
- *The role of utility companies.* Utility companies have become one of the largest drivers of fiber-to-the-home connectivity, using their existing pipes and ducts to enable faster, less expensive deployment. In Norway, for example, utilities now connect more than 6 million citizens with wireline broadband access. Service providers must therefore consider what role utility companies might play in the overall high-speed broadband ecosystem.
- *Potential savings*: High-speed networks can be less expensive to operate. A carrier's real estate portfolio, for example, will be substantially smaller along with its investment. Higher capacity optical switching requires less floor space in exchanges and can provide greater reach from exchanges to customer premises. The lower energy cost of an all-fiber network also contributes to savings. These savings will not materialize, however, if providers must maintain both their new broadband network and their copper legacy network.[13]
- The network deployment is focusing on the most profitable geographical areas, primarily major urban areas and intercity routes. They create an imbalance between towns and rural areas, paving the way for a divide which might keep the rural inhabitants out of the stream.

The success story of mobile phone companies in Africa is well known and documented. However, lessons learned from its flaws do not seem to enlighten the decision-making process related to broadband deployment. There is a real danger of serious policy mistakes. As in developed markets, broadband strategies in developing countries tend to focus on investment in optic fiber. The focus on fiber may miss the transformational opportunity provided by broadband, video, and the unmatched content of Africa.

Accenture: Reshaping the business for sustainable digital growth, Why a new operating model is needed for high performance in tomorrow's digital Media and Entertainment industry. Consulted May, 10, 2012.

[13] Ibid.

References

Perez, S. (2007). Economic effects of increased broadband use in California research report.
Williams, M. D. J. (2010). *Broadband for Africa, developing backbone communications networks*, Infodev, The World Bank.
Gunasekaran, V. (2005). Migration to 4g- .ubiquitous broadband.economic modeling of wi-fi with wimax. Stevens Institute of Technology Telecommunications Management (Castle Point on Hudson)
Restarick, J. C., & D'Aleo, M. (2009). Planning an effective high-speed broadband deployment. Retrieved from http://www.accenture.com/SiteCollectionDocuments/PDF/PlanningBroadband_10509.pdf.
The Economist. (2009, September 24). Mobile marvels.
www.accenture.com/. . ./insight-bringing-tv-life-race-dominate.
ITU The world in 2010: ICT facts and figures.
ITU StatShot – Issue 8, January 2012.
Khalil, M. A. (2010). *Broadband for Africa. Developing backbone communications networks.* Infodev, The World Bank Group.
Akpan-Obong, P., & Alozie, N. O. (2011). *Broadband technology and development: Access critical to African economies?* Mesa, AZ, USA: Arizona State University, Polytechnic Campus.
In Alison Gilwald, A. (Eds.). (2010). *Towards an African e-Index: Household and individual ICT access and usage across 17 African countries.*

Bibliography

Abiola, O. E. (1978). *History of West Africa.* Ado-Ekiti: Ondo State Printing and Publishing House.
Ake, C. (1982). *Social science as imperialism.* Ibadan: Ibadan University Press.
Anikulapo-Kuti, F. (2011, July 4). Fela in Berlin. Jazz festival of 1978 by the AfroBeat King. YouTube.
Al-Aswany, A. (2011). *On the state of Egypt: A novelist's provocative reflections.* Cairo: The American University in Cairo Press.
Arthur, A. (2007). Global communication culture: A challenge to spirituality of the heart. Retrieved June 13, 2011, from www.misacor.org.au.
Encarta Encyclopaedia. (2011). *Microsoft Corporate Software.* New York.
Hafez, K. (2007). *The myth of media globalization.* Cambridge: Polity Press.
Henley, A., & Schott, J. (2002). *Culture, religion and patient care in a multi-ethnic society.* London: Age Concern England.
Lassiter, J. E. (2011). African culture and personality: Bad social science, effective social activism, or a call to reinvent ethnology?, *African Studies Quarterly, The Online Journal for African Studies*, Retrieved June 13, from www.africa.ufl.edu/asq.
Louw, E. (2008). *The media and political process.* London: Sage Publications Ltd.
MaCBride, S., et al. (1981). *Many voices, one world.* Ibadan: Ibadan University Press.
Markovitz, I. L. (1969). Liopold Sedar Senghor and the politics of negritude. Dakar: Abe Books.
Mazrui, A. A., & Mazrui, A. M. (1995). *Swahili state and society: The political economy of an African language.* Nairobi: East African Educational Publishers.

McLuhan, M. (1964/1995). *Understanding media: The extensions of man*. London: Ruthledge and Kogan Paul.
Mehrabian, A. (1981). *Silent messages* (2nd ed.). Belmont, CA: Wadsworth.
Omu, F. I. A. (1978). *Press and politics in Nigeria: 1880–1937*. London: Longman Group Limited.
Osho, S. A. (2010). *African communications system*. Abeokuta: ESS-OH Consult Publications.
Rodney, W. (2009). *How Europe underdeveloped Africa*. Abuja: Pearsons Publishing Co. Limited.
Scannell, P. (2009). *Media and communication*. London: Sage.
Schiller, H. (1976) Communication and Cultural Domination. New York: International Arts and Science Press.
Soyinka, W. (2011, July 2). Biography of Wole Soyinka. www.wolesoyinka.com.
Wikipedia. (2011). Free encyclopaedia. Online.
Wilson, D. (1999). Traditional media in modern Africa development. *African Media Review. 1*

Chapter 7
Broadband as a Platform for Video Delivery: What to expect from Platforms and Applications

Brian Arendse, Adedamola Adedeji, Mlungisi Mhlungu, George Thomas, Shiv Bakhshi, and Sadiq Malik

Introduction

Video consumer behaviour is changing. While broadcast TV is still popular for news and live events, consumers use a variety of platforms and different ways to view content. Video is an enabler for business, society and improving literacy in Africa. There is a shift towards media services that focus on the individual, are simple to use and deliver on-demand content in a way that meets user expectations for quality (Ericsson 2011).

The challenge to delivery of video content is the requirement for content delivery to multiple screens efficiently. Multiple screens include television, tablets, personal computers (PC) and mobile devices like smartphones. Centralised content distribution mechanisms and platforms are required to efficiently deliver video content to multiple devices. The availability of high-speed broadband technologies like LTE is making video delivery a reality in Africa.

This chapter outlines technology and systems that could enable efficient delivery of video content to multiple devices using broadband as the primary medium for delivery of the content. Three topics to be covered in this chapter include:

1. Broadband market overview including consumer trends
2. Network evolution to support video
3. Efficient content delivery

B. Arendse (✉)
Ericsson Region sub-Saharan Africa, 148 Kelvin Drive, Woodmead, South Africa
e-mail: brian.arendse@ericsson.com

J. O'Neill et al. (eds.), *Broadband as a Video Platform*, The Economics of Information, Communication, and Entertainment, DOI 10.1007/978-3-319-03617-5_7,

Broadband Market and Consumer Trend Overview

Video: The Biggest Driver for Internet Traffic Growth

Global IP traffic grew 70 % year on year during 2012 to 855 petabytes a month with the highest growth rates experienced by the Middle East and Africa (101 %) and Asia Pacific (95 %). The average monthly traffic in 2014 will be equivalent to 32 million people streaming a high-definition 3D movie, continuously for the entire month. Global IP traffic will quadruple from 2009 to 2014. Overall, IP traffic will grow at a compound annual growth rate (CAGR) of 34 %.

It would take over 2 years to watch the amount of video that will cross global IP networks every second in 2014. It would take 72 million years to watch the amount of video that will cross global IP networks during calendar year 2014.

The sum of all forms of video (TV, video on demand (VoD), the Internet and P2P) will continue to exceed 91 % of global consumer traffic by 2014. Internet video alone will account for 57 % of all consumer Internet traffic in 2014. By 2014, 3D and HD Internet video will comprise 46 % of consumer Internet video traffic.

Video communications traffic growth is accelerating. Though still a small fraction of overall Internet traffic, video over instant messaging and video calling are experiencing high growth. Video communications traffic will increase sevenfold from 2009 to 2014. By 2014, Internet TV will be over 8 % of consumer Internet traffic, and ambient video will be an additional 5 % of consumer Internet traffic (Cisco 2013).

Important broadband market trends being seen globally include the following:

- Globally, data traffic exceeded voice traffic in 2010, and by 2014, 90 % of data traffic will be on mobile networks.
- Video is driving mobile traffic growth—by 2014, video will account for 66 % of global mobile data traffic.
- User expectations are rising—end users will click away or abandon if website or video fails to load within a short few seconds.

Within the concept of mobile broadband communications, video is seen as the new voice with online video being the biggest contributor to traffic volumes (Ericsson Mobility Report 2012). Figure 7.1 shows the dominance of video compared to other broadband data traffic.

Changing Consumer Media Consumption Patterns

Video Snacks

Video snacks are a new form of media consumption evolving in parallel with the social networking phenomenon. In general, video snacks are of short and low

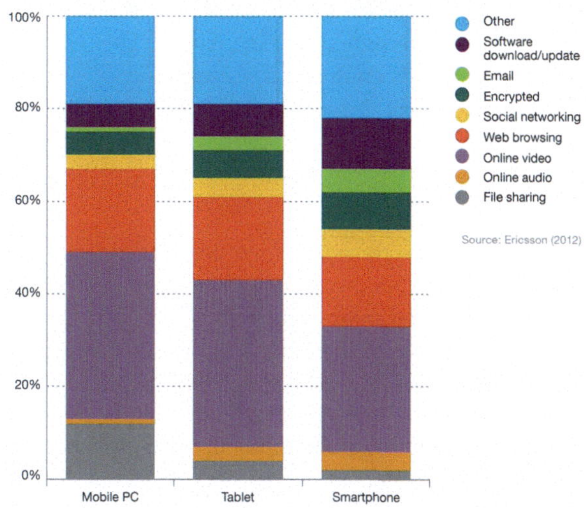

Fig. 7.1 Application Internet traffic volume by device type (Ericsson Mobility Report 2012)

resolution. They are extremely inexpensive to produce or have been appropriated completely free of charge by the person who wishes to share them over the Internet.

Internet Television

Internet television is a television-like experience over the public Internet. You can find examples of Internet television at most prominent broadcasters and content providers for example DSTV and fox.com. Almost every major television network offers some kind of online viewing experience for their most popular shows. This video content does not have a name that everyone agrees upon. Most people just call it "broadband video" or "online video".

Download to Own

Download to own is a well-understood consumer behaviour. iTunes, Amazon-On-Demand and other download-to-own (or rent) services operate under the assumption that consumers will be willing to pay for files and to download, store and then view them. Of course, digital files are hard to protect from piracy. People with more money than time buy, and people with more time than money pirate.

Video on Demand

Whether it is video snacking, Internet television or download to own, recent observations of the changes in video consumption lead to one overwhelming conclusion—consumers like to consume video content and they like to be in control of their media consumption experiences. This consumer trend shows significant inclination towards on-demand consumption.

Video Applications

It is important to consider video applications that are also having a significant impact on consumers (webcams) and businesses (video conferencing). As travel becomes more and more difficult and expensive, you can expect two-way video applications to approach ubiquity in broadband environments.

Video is everywhere, and it is being produced and consumed at record levels. Video is becoming more personal, more interactive and more social. Both production and consumption are trending upwards and will continue to do so indefinitely.

Network Evolution to Support Video Delivery

Broadband Technology Overview

Broadband can be delivered in various ways, fixed and mobile for different applications with each technology delivering various speeds. The fixed access broadband uses technologies such as ADSL and fibre. We see fibre to the home delivering as much as 1 Gbps for example.

The focus for Africa will be on mobile: GSM/GPRS/EDGE, HSPA and LTE. We see wireless technologies such as LTE also delivering as much as 1 Gbps on the air interface. EDGE/Evolved EDGE and HSPA have also been improved to deliver speeds in the range of Mbps. So, the technology is no longer a limiting factor as we see high speeds also being supported on wireless technologies. Actual end-user throughputs may be up to ten times less than the theoretical speeds shown in Fig. 7.2 due to radio conditions, signal strength and network loading or capacity constraints.

Mobile Broadband Adoption in Sub-Saharan Africa

When looking at broadband technology deployment in sub-Saharan Africa, mobile broadband based on WCDMA/HSPA and LTE will be the technology of choice for delivery of broadband. These deployments will be dependent on the deployment of terrestrial fibre for backhaul. Figure 7.3 shows 100 % WCDMA launches by 2013 in most countries in sub-Saharan Africa. The commercial launch of LTE services in

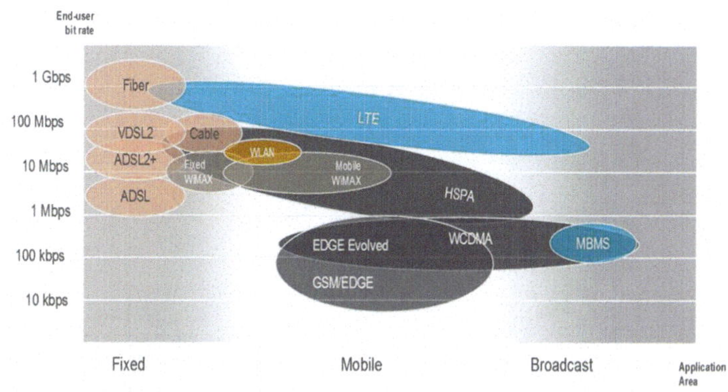

Fig. 7.2 Broadband network technology overview

sub-Saharan Africa will happen as spectrum becomes available and as the business case surrounding it becomes viable.

Angola, Namibia and South Africa are early adopters of LTE.

One TV, Many Devices: Data Offload Strategies Using Wi-Fi

The average home entertainment set-up in more economically advanced countries in sub-Saharan Africa is moving away from using separate screens in each room to video services supplemented by a number of mobile devices that provide access to these services from all over the home. Tablets in particular are growing into a popular device for viewing content in the home. This is partly due to the favourable conditions for mobile viewing within the home and the fact that many consumers of video content are looking to modernise their TV experience yet prefer not to invest in more than one new TV or add set-top boxes to their old TV sets. On-demand services that cater for more focused viewing situations, usually on impulse, are gradually changing the way of watching video (Ericsson ConsumerLab 2012).

Delivering efficient radio network capacity and coverage to cater for this increased demand for video services is central to most mobile operators' mobile broadband strategies, and Wi-Fi is a key element to satisfying the data appetites of a growing number of smartphone and tablet users in sub-Saharan Africa. With Wi-Fi fully integrated into mobile access and core networks—offering seamless, secure roaming, intelligent radio access-type selection, mobility and carrier-grade scalability and manageability—operators will be able to optimise the user experience and take advantage of a variety of flexible new business offerings. The new market reality of an all-you-can-eat wire-grade fixed broadband experience at home and office offers both a challenge and an opportunity to mobile operators who are positioning themselves to bring Wi-Fi access into the telecom mainstream and make the end user experience as convenient, responsive and seamless as possible.

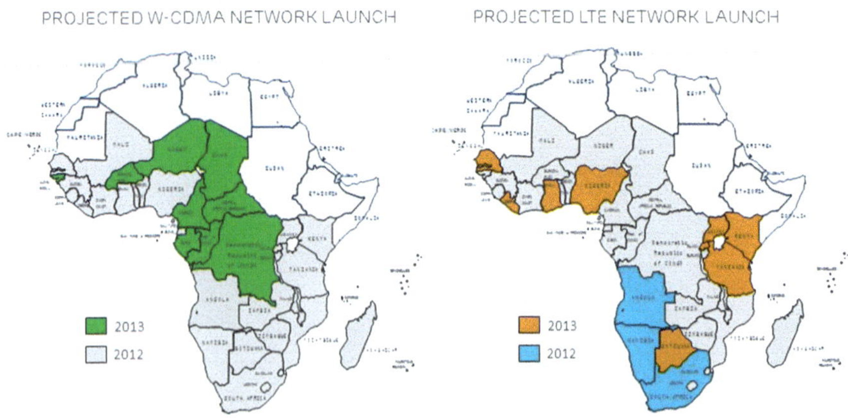

Fig. 7.3 Mobile broadband data launch in sub-Saharan Africa 2012–2013 (Informa World Cellular Information Service 2013)

In typical high-traffic areas where deploying network resources is more of a challenge (for example, in homes with high video demand or public areas such as airports and hotel lobbies), it may not be feasible to densify or improve the macro network within specific time, cost or spectrum constraints. Operators can then add small, low-power cells to make use of both licensed and unlicensed bands, including the rollout of network-integrated Wi-Fi (Ericsson Whitepaper 2012a). With solutions that are scalable and integrated into the core network and that enable good visibility and management of the user experience, Wi-Fi can be used to deliver all the same services available from the cellular data network, maximising the entire user experience by delivering consistently high-performance broadband.

A seamless user experience requires end-to-end integration all the way from the mobile packet core network to individual cell or access point, encompassing controllers and management systems along the way, as illustrated in Fig. 7.4.

With Wi-Fi supporting seamless integration with both existing mobile core and fixed edge infrastructure, operators can leverage existing assets to improve the user experience while reducing their total cost of ownership (TCO).

Service Provider Strategy Around Video Delivery

The consumer media consumption paradigm is changing at an ever-increasing pace. And the Internet is evolving into a media-centric environment. However, on the service provider's (SP) side, the media distribution paradigm is not changing anywhere near as quickly. This supply and demand mismatch is responsible for much of the angst SPs feel as they try to compete for today's media-hungry consumers. The good news is that most SPs across the globe have taken the step to upgrade their network infrastructure to next-generation networks (NGN).

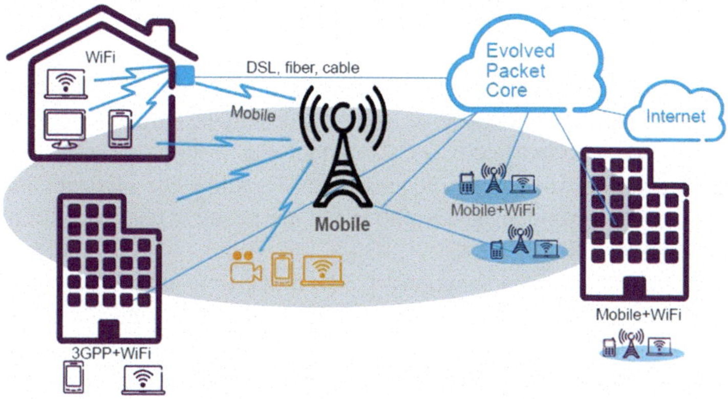

Fig. 7.4 Wi-Fi access network integrated into the mobile packet core for efficient video broadband delivery (Ericsson Whitepaper 2012a)

CSPs know that their infrastructure has to be ready to meet bandwidth demands and that their service offerings must be bundled to reduce churn and increase revenue. Key insights into the future of profitable video distribution can be gained by studying advertising efficacy on Internet television. Publishers have demonstrated that you can accomplish a great deal by creating scarcity (reducing inventory) and increasing relevance by using full-episode Internet television players. The next logical step is the ability to deliver content to a targeted audience at a premium price. It will be a fundamental building block of any future service offering.

In a media-centric broadband world, service offerings will have to include more than on-demand content. Although one could fantasise about a lot of new, interesting ways to productise media, service providers don't have to work that hard. The best ROI is more likely to come from adding small conveniences that help the user experience. The importance of the "cool" factor cannot be overstated. If the experience is remarkable, consumers will consume.

Telcos know that their infrastructure has to be ready to meet bandwidth demands and that their service offerings must be bundled to reduce churn and increase revenue. The technology for a fully integrated multi-platform system exists. It just needs to be productised. Right now, there are literally hundreds of different products and services that are very appealing to consumers. They create a great deal of value, but so far, very few organisations have been successful in converting that value into wealth.

Efficient Video Content Delivery

Media delivery networks (MDNs) enable efficient delivery of media like video. Content delivery networks (CDNs) enable the efficient delivery of all IP traffic through agreements with content providers.

Media Delivery Networks

An MDN helps operators cope with the growth of video content. The media content equation contains three parameters: content consumers, content providers and content delivery. All three elements affect the way technology will develop to create a universal cost-beneficial media system. Handling the massive amount of over-the-top (OTT) or unmanaged traffic is a major business challenge for network operators (Ericsson 2011).

MDN includes three core functions: CDN, transparent Internet caching (TIC) and service and performance enhancers (S&PEs), shown in Fig. 7.5. In the MDN set-up, the network is enabled for delivery of video as opposed to traditional delivery of voice and data with video being prioritised over other data services allowing for improved network quality of service (QoS).

Local caching and adaptive streaming improves the end-user quality of experience (QoE) for delivery to multiple screens. MDN enables increased traffic and improved operator revenues. Network optimisation including compression, caching and streaming allows operators to decrease operational expenses.

Three potential MDN business models or use cases are shown in Fig. 7.6. In the "build" or operator-managed model, the operator owns or sources content and delivers the content directly to its subscribers. In the wholesale network model, one entity secures content from content providers, a managed service platform is set up and media distribution capacity can be sold directly to end users or to other service providers for further re-distribution to end users.

In the OTT services caching use case, the MDN's TIC capabilities allow the operator to minimise the cost of delivering OTT services. TIC is transparent to both the content provider and the consumer of the service. TIC helps the operator to significantly reduce peering and transit costs as well as reduce the need to invest in internal network upgrades to support growing volumes of OTT traffic.

Telcos have access to a whole host of customer data. They can supply information on what people have previously downloaded and when (day and time), combined with location information and possibly presence information as well. This could provide a powerful tool to proactively push information about video options to consumers even before they're aware that they might want them but at times when they are receptive to them. Such functionality also needs to be backed by an immediate and simple payment and access system so that the impulse to buy can be taken instantly to its natural end.

Content Delivery Networks

Internet traffic is growing exponentially and continues to be dominated by video. As customers drive the demand for this disruptive trend, operators are forced to consider innovative solutions to manage the rapid growth in managed and

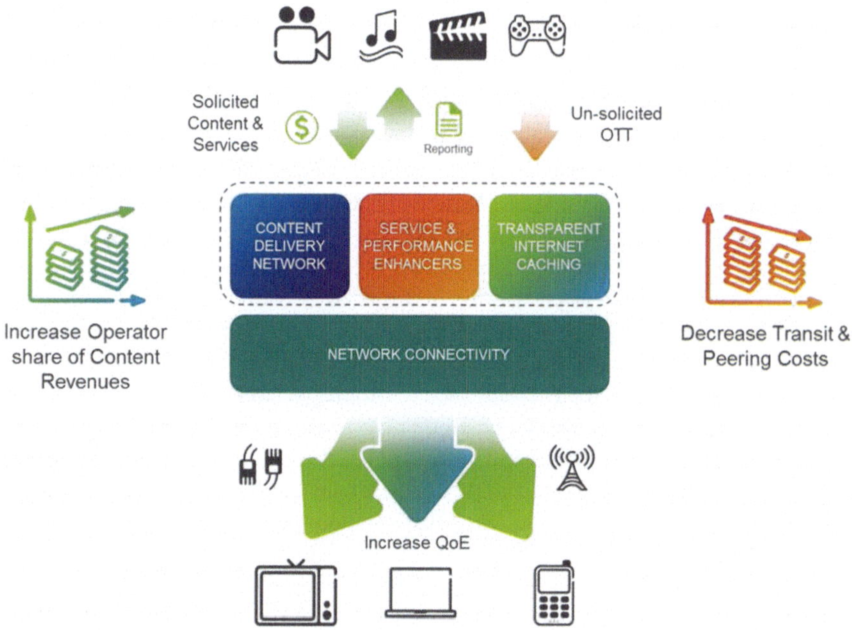

Fig. 7.5 Media delivery network solution

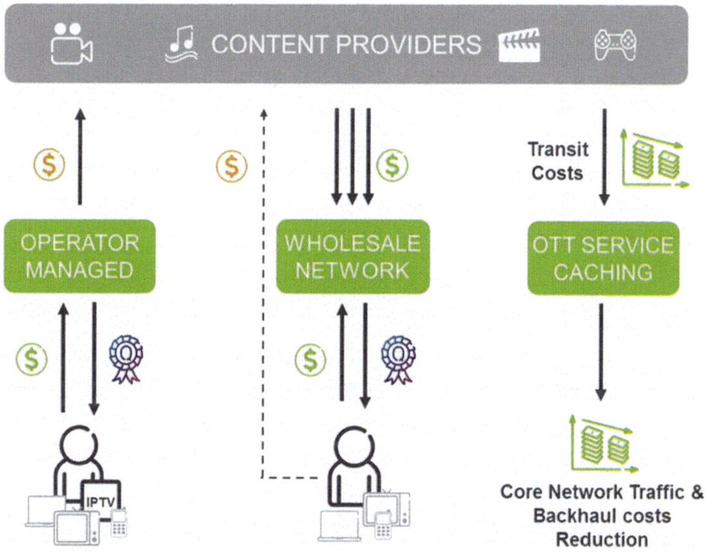

Fig. 7.6 Media delivery network business models and use cases

unmanaged content, especially the massive growth of OTT services and applications. These services and applications can present network operators with several challenges including high peering costs, rising backhaul transit costs, mounting last-mile bandwidth demands and decreasing subscription revenues. A CDN is a service provider solution that enables the distribution of content based on agreements with the content owners across the Internet as far as an exchange point or in some cases further into an operator's network when a cooperative agreement has been reached (Ericsson Whitepaper 2012b).

This system is enabled by positioning Internet content servers as close to the end user as possible—either physically or in terms of network topology (number of hops). This solution represents a significant shift in investment focus to OPEX, enabling operators to enter the media value chain with profitable video delivery, thereby leveraging off their established customer relationships while offering content providers and enterprises cost-effective accessibility and guaranteed QoE across mobile and fixed networks. The benefits of this solution include innovative, user-aware video optimisation and delivery saving bandwidth, greater usage of network capacity and richer video broadband experiences.

Everyone can enjoy the benefits of this model—users get access to content and content providers gain a means of global distribution. The main problem though relates to delivery, which arises between the provider and the consumer. The root cause of the problem lies in the fact that the delivery model relies on the core Internet—designed for robustness rather and providing guaranteed QoS—to get content from one place to another. This creates bottlenecks both in terms of bandwidth and latency, as illustrated in Fig. 7.7.

Efficient Internet content delivery of services such as broadband video can be achieved through effective content distribution technology aimed at reducing bandwidth costs for operators. Several concepts such as traffic redirection, caching, multi-protocol delivery and content migration can be applied to both fixed and mobile network architectures to deliver optimised Internet-based content, enabling mobile, fixed or converged operators to create new revenue streams via wholesale content distribution and delivery, differentiate their services through S&PEs and reduce operational costs and capital expenditure. External content providers and operators can benefit from optimally designed CDNs through improved delivery of content through the operator's enhanced delivery capacity and through increased end-user retention due to improved QoS (Ericsson 2011). The functional architecture of an efficient CDN is illustrated in Fig. 7.8.

Efficient Video Delivery to Multiple Screens

The challenge to delivery of video content is the requirement for content delivery to multiple screens efficiently. Centralised content distribution mechanisms and platforms are required to efficiently deliver video content to multiple devices.

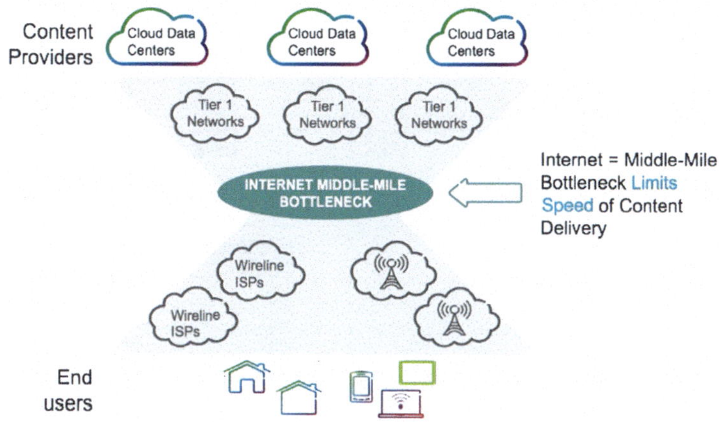

Fig. 7.7 Problems with Internet delivery of content to end users

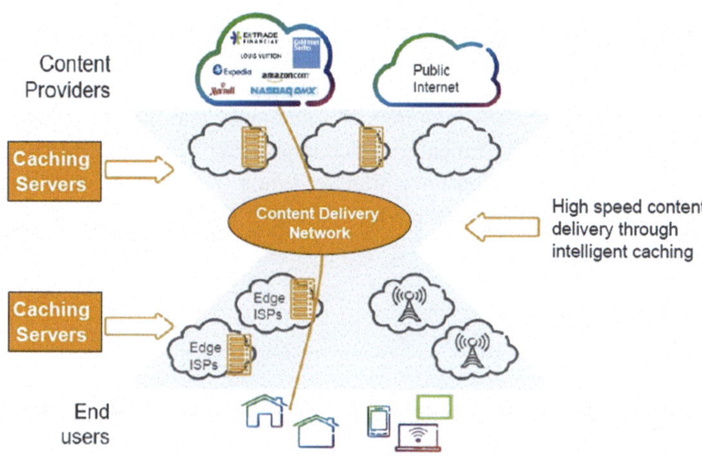

Fig. 7.8 Efficient Internet content delivery architecture through CDN and caching servers

Linear video is delivered to television by terrestrial broadcast, satellite or IP-TV. The quality requirement set by the screen size and resolution determines the video bit rate. Adaptive streaming works by adapting the video stream bit rate in real time to the actual network throughput to a given endpoint without the need for "re-buffering". If the network throughput suddenly drops, the picture may degrade but the end user still sees a picture. Apple's HTTP Live Streaming (HLS), Microsoft Silverlight Smooth Streaming (MSS) and Adobe's HTTP Dynamic Streaming (HDS) are examples of HTTP adaptive streaming solution implementations.

Adaptive streaming makes changes at the server and the client to increase the overall QoE of the end user. To support adaptive streaming, the content is first encoded at multiple bit rates and resolutions which must be predefined by the

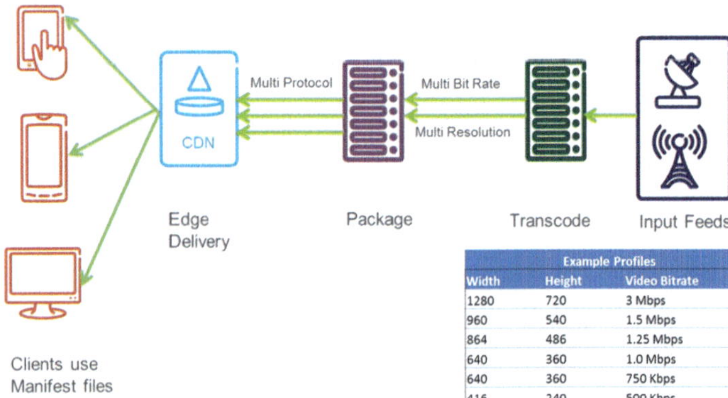

Example Profiles		
Width	Height	Video Bitrate
1280	720	3 Mbps
960	540	1.5 Mbps
864	486	1.25 Mbps
640	360	1.0 Mbps
640	360	750 Kbps
416	240	500 Kbps
320	180	350 Kbps
320	180	150 Kbps

Fig. 7.9 Video content over IP delivery using adaptive HTTP streaming

operator to provide an acceptable trade-off between quality and bit rate. Segmentation or packaging allows a client to switch from one stream to another seamlessly. The adaptive part of adaptive streaming is enabled at the client rather than the server. The client continually monitors the available bandwidth and the media being delivered and will dynamically switch to a higher or a lower bit rate session in order to keep the receive buffer within set limits (Adams 2013). A high-level implementation of an adaptive streaming solution is shown in Fig. 7.9.

Conclusions

Research and data indicate that video traffic is the biggest contributor to mobile IP traffic. End users require multiscreen video capability (TV, PC, smartphone, tablet).

Broadband platforms in the form of fibre and mobile technology are making video delivery for business and personal applications a reality in Africa. Mobile broadband is becoming a reality in Africa with 3G launches planned in all sub-Saharan African countries in 2013 and LTE deployments following soon thereafter.

Different business models exist for media delivery enabling efficient and cost-effective solutions for service providers and content delivery solutions with caching and QoS features existing for efficient video delivery. Carriers are planning to integrate IPTV, mobile TV and Web TV/video offerings into a seamless user experience in and out of the home topped by new value-added services.

Technology convergence in media, new media offerings and ways to deliver content from the Internet are fast emerging as a sound way to distribute and consume content online. As such, Telcos/CSPs need to adopt innovative business models in B2B, B2C and OTT scenarios to boost revenue streams as well as understand how to shape up the media consumption model and move up the media business value chain. The new media offerings and ways to deliver content OTT are fast emerging as a sound way to distribute and consume content online.

The rollout of next-generation broadband networks will increase competition in the supply of video content into the home. The manner in which the Internet and fibre infrastructure will be a "universal cross-connect" for the delivery of digital video content into the home must be understood. The vast amount of content (video, audio, photos, etc.) suggests that a new way to discover media and personalise the experience is needed.

When mobile and telco carriers decide to enter the video business, they need a distribution architecture that can handle change at unprecedented price points based on a solid business case. Constructing a viable business case with the use of various broadband technologies, commercial objectives and project managing a successful rollout will be keys to success.

References

Adams, M. (2013). *Will HTTP adaptive streaming become the dominant mode of video delivery in cable networks?*, Retrieved April 1, 2013, from http://archive.ericsson.net/service/internet/picov/get?DocNo=28701-FGD101097&Lang=EN&HighestFree=Y

Cisco. (2013). *Cisco visual networking index: forecast and methodology, 2011–2016 White Paper*. Retrieved 29 March, 2013, from http://www.cisco.com/en/US/solutions/collateral/ns341/ns525/ns537/ns705/ns827/white_paper_c11-481360.pdf

Ericsson. (2011). Managing the growth of video over IP. *Ericsson Review, 1*. Retrieved 18 February, 2013, from http://www.ericsson.com/res/thecompany/docs/publications/ericsson_review/2011/Media-Delivery-Network.pdf

Ericsson ConsumerLab. (2012). *TV and video—an analysis of evolving consumer habits*. Retrieved 15 February, 2013, from http://www.ericsson.com/res/docs/2012/consumerlab/tv_video_consumerlab_report.pdf

Ericsson Mobility Report. (2012). *On the pulse of the networked society*. Retrieved 18 March, 2013, from http://www.ericsson.com/res/docs/2012/ericsson-mobility-report-november-2012.pdf

Ericsson Whitepaper. (2012). *Wi-Fi in heterogeneous networks—a seamless transition*. Retrieved 18 February, 2013, from http://www.ericsson.com/res/docs/whitepapers/wp-wi-fi-in-heterogeneous-networks.pdf

Ericsson Whitepaper. (2012). *Operator opportunities in cloud service delivery*. Retrieved 18 February, 2013, from http://www.ericsson.com/res/docs/whitepapers/wp-cloud-opportunities.pdf

Informa World Cellular Information Service, WCIS. (2013). Retrieved 18 March 2013, from http://www.informatandm.com/about/wcis/

Chapter 8
Technology Trends to Connect Africa

Martin Phelps

Introduction

This chapter is drafted from the perspective of the large multinationals that are AT&T's focus outside the USA and which are driving much of the inward investment that Africa has been benefitting from in recent years.

In common with many network service providers, the portfolio that AT&T offers in Africa is a subset of what it offers elsewhere in the world. Lack of ubiquitous broadband is one of the restrictions currently faced. Thus, this chapter is very much a forward-looking view of what could be achieved in Africa through a concerted effort by all the private and public stakeholders to implement an environment that fosters growth and innovation.

A Connected World

The connection between investment in Information and Communication Technology (ICT) and economic development is well understood and, indeed, the underlying purpose of the conference that this chapter, in part, memorialises. However, the diagram below makes clear that global trade flows have developed predominantly along the main global communication "corridors". That is, there is an extremely high correlation of inter-country/regional trade and the degree of communication capacity.

M. Phelps (✉)
AT&T Business Solutions, London, UK
e-mail: mp8564@emea.att.com

J. O'Neill et al. (eds.), *Broadband as a Video Platform*, The Economics of Information, Communication, and Entertainment, DOI 10.1007/978-3-319-03617-5_8,

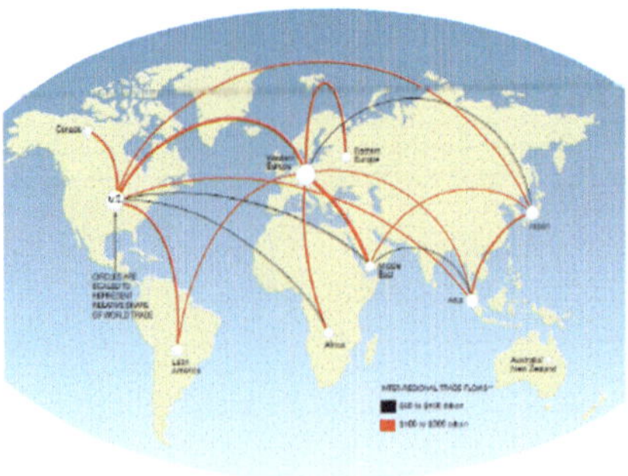

Trade flows - 1980s

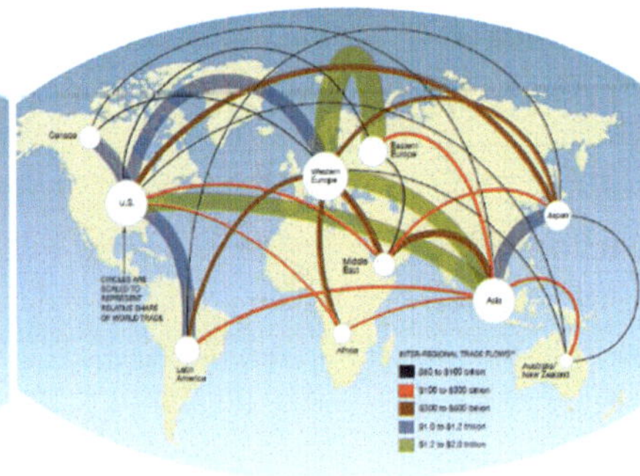

Trade flows - Late 2000s

Technology Driving Economic Growth

Research that AT&T commissioned in Europe with Oxford Economics on business productivity further underscores the important link between trade and growth and ICT investment. The chart, below, illustrates the impact of ICT on labour productivity and in particular that ICT's impact in the USA exceeds the total labour productivity growth in Belgium/the Netherlands and Italy, and almost exceeds that in Japan:

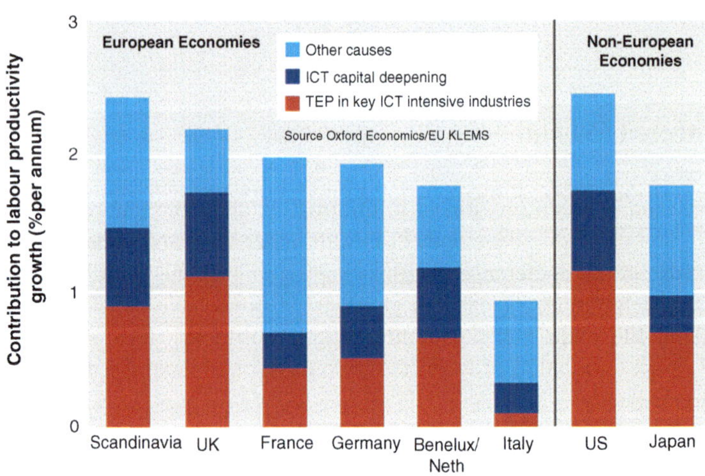

- Capital deepening refers to improvements to a firm's "capital stock"—the entirety of a firm's assets, including assets like software and intellectual property.

- TFP is "total factor productivity" and includes technical, know-how, improvements in business organisation and managerial efforts that reduce total costs.

Clearly some countries see better productivity returns from ICT investment than others. As seen above, the European productivity leaders are Scandinavia and the UK, which have invested the most among their regional peers in ICT and have market conditions favourable to exploit it. Over the past 15 years, they have seen average labour productivity growth of between 1.7 and 2 % a year.

But both of them and the rest of Europe have fallen behind the world leader in investment in ICT—the USA—since 1991. The USA increased its accumulated stock of ICT investment as a proportion of GDP from 9 % in 1991 to 30 % in 2010. Europe's ICT capital stock increased from 6 to 9 % (near parity with the USA) to around 20 % over the same time frame.

The research shows that European GDP could grow by an additional €760 billion, or an extra 5% above forecasts, if Europe matched total US ICT levels by 2020.

This would be worth around €1,500 per person at today's prices. ICT-driven innovation would contribute approximately one-third of that growth—1.5% of GDP or around €220 billion.

Another example of how technology drives growth comes from the US wireless industry:

A study by Recon Analytics, authored on behalf of CTIA, quantifies the wireless broadband industry's impact on the US economy. Analysis finds that every 10 MHz of additional licensed spectrum allocated for commercial use will result in $1.7 billion in GDP and 7,000 additional US jobs.

The industry added $195.5 billion to the US GDP between July 2010 and June 2011. If that amount were equivalent to a nation's GDP, it would be the 46th ranked economy in the world. In 2011, wireless services produced $33 billion in productivity improvements for the US businesses in nine categories. Over the next decade, productivity gains attributable to wireless are expected to total more than $1.4 trillion.

Service Strategies Being Adapted by Enterprises to Leverage Technological Advances

So the question becomes the following: What do organisations, both public and private, have to do to be able to harness these growth drivers and productivity opportunities?

AT&T's customers tell us that the three main demand drivers in the large enterprise segment today are mobility, globalisation and virtualisation (improving cost, labour flexibility and security through a process of abstracting computing resources to a separate environment, remote from where the applications are used).

Technology can help to address these demand drivers, and so technology becomes the platform for delivering growth.

One great example is the combination of high-speed mobile broadband and virtualisation—often referred to as "cloud" services.

We are starting to see applications where the content that might previously have been held on a smartphone or a tablet PC is being hosted in the cloud. Once the content is "virtualised" in this way it is accessible from any device with the speed to download it. That is one of the reasons that the applications are starting to be available now, as 4G high-speed mobile broadband, such as LTE, makes possible technological improvements over the next 3 or 4 years that are fundamentally going to change the value equation in the ecosystem.

Video Is Becoming a Mainstream Collaboration Technology

There is no doubt that video is a key application to drive productivity. As we all know, video technology and videoconferencing are not new. However, video has been the subject of a great deal of attention and analysis recently, whereas it was slower to gain traction in the market in the past.

There are two primary reasons why video is succeeding now, when it failed in the past. The first is that unified communications (UC) technology has allowed video to be integrated with presence, email, messaging, etc. to bring a multiplier effect to its productivity benefit.

Secondly, the new video systems are a lot better than their predecessors. Today's immersive telepresence systems, with life-size participants in high-definition quality video, are a giant leap forward in simulating the face-to-face experience.

However, such systems are quite expensive, and their uses are somewhat limited outside of communities whose interactions can justify the cost. So we are starting to see the introduction of desktop telepresence systems, at a much lower price point, that will drive a further wave of productivity gain when there is mass adoption; and that will enable a much broader range of activities to benefit from collaboration technology. Many of those activities or applications were discussed at the Zambia conference in 2012 and are memorialised in the chapters in this volume.

Transforming Business Using Unified Collaboration and Communications

The diagram, below, illustrates how multiple communication and collaboration tools can be brought together and managed from a single desktop client, in this case the AT&T UC client.

Having a single interface to all your UC applications enables integration of those services with critical business processes and core applications.

It is called communications-enabling a business process to:

- Reduce communication delays
- Optimise business processes for efficiency
- Accelerate the pace of business
- Improve the ability to respond to key issues

Forrester Research identified six ways a firm can benefit by communications-enabling business processes to:

- Provide faster access to experts
  - Example: Adding a "click to chat" feature to your support website
- Shorten the sales cycle
  - Example: Automated notifications to "ping" stakeholders during contract approval process
- Reduce travel costs
  - Example: Collaborate across geographies with a video conference
- Improve customer experience
  - Example: Contact centre agents with presence info on experts to increase first-call resolution rates, satisfy customers
- Rapid problem resolution
  - Example: Present information to locate nearest inspector before quality issues impact shipping dates
- Accelerate projects

– Example: Bring engineers together to collaborate online during design project

UC goes beyond increasing personal efficiency to increasing business process efficiency.

Technology Driving Social Benefit: Healthcare Solutions

In the USA AT&T has formed a dedicated "eHealth" organisation and developed a modular and incremental approach to adopting health information technology that gives doctors and other providers access to best-of-breed applications, including ePrescribing, telemedicine, picture archiving and communication system (PACS) images, lab results, radiology reports and others. The multivendor approach lets healthcare providers share information easily and securely. mHealth refers to a series of services that provide automated clinical and behavioural coaching for patients and their caregivers, such as DiabetesManager.

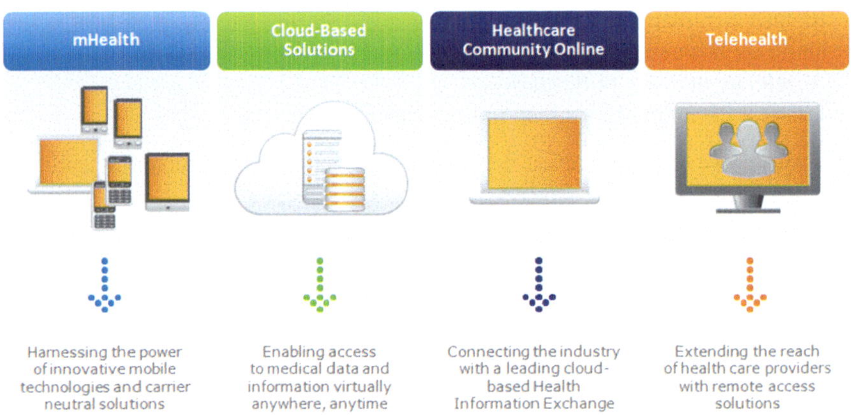

Physicians with access to electronic medical records can immediately view anyone's medical history, enabling faster and more complete diagnoses. Making records available online reduces the time spent tracking down and sharing hard copy medical records and eliminates the need for duplicative medical tests, which reduces healthcare costs. Health information exchanges help minimise costly mistakes, giving physicians all relevant information about a patient, including, for example, allergies to certain types of medicine.

Electronic medical records are becoming a vital part of patient care with the development of regional HIEs like the one used in Memphis. Dozens are springing up across the country, representing the leading edge of the growing eHealth movement, which seeks to use information technology to improve the quality and reduce the cost of healthcare.

It is not just medical records that have become electronic. Another common eHealth initiative is ePrescribing, which lets physicians use the Internet and other online tools to create and sign prescriptions electronically. eHealth experts say that ePrescribing can reduce medical errors, decrease pharmacy costs and increase efficiency, in part by giving prescribers real-time information about potential drug interactions and allergies and eliminating mistakes associated with the interpretation of handwritten prescriptions. ePrescribing reduces the need for phone calls between doctors and pharmacists and provides patients with safer care.

Telemedicine is a third eHealth tool that takes advantage of telecommunications technology to enable long-distance patient consultations and delivery of clinical diagnosis and other medical services. After the devastating earthquake in Haiti, physicians at the University of Miami's temporary tent hospital in Port-au-Prince used teleconferencing and amateur radio transmissions to consult with specialists back in Miami and other US locations as they treated thousands of earthquake victims. Even when there is no disaster, telemedicine reduces the need for travel by patients and physicians, facilitates faster and more consistent treatment of injuries and illnesses and allows some patients to be treated at home rather than in a hospital. It also leverages the benefits and skills contained in a medical facility in one city of a country to be taken advantage of by patients in other cities of the country where the patient docs not have the financial resources or time to travel to the central hospital.

Greater Reliance on Technology Requires Greater Diligence to Secure the Underlying Systems, Applications and Data

Today's Internet security threats range from curious prowlers to savvy intruders, simple mischief to espionage. Without a plan to help protect your entire network and its connection points, defence is only as strong as its weakest link. We have started to become de-sensitised to the constant stream of security breaches and consequent financial loss, but the attack on Sony's PlayStation Network in 2011 was estimated to have cost the company about US$170 million.

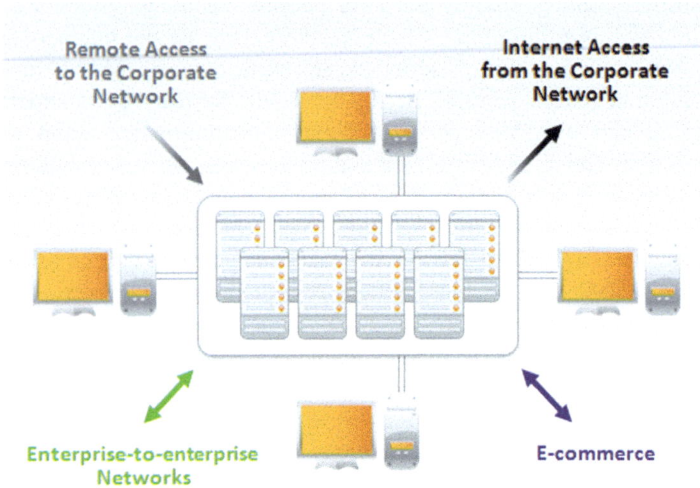

AT&T uses "defence in depth" security architecture, with security features built into every network layer and every supporting process. The theory of "defence in depth" is based on the concept that multiple diverse security measures are intrinsically more effective than a single homogenous defence. So, if the security measures in the first network interface layer are breached, security measures placed inside the network edge at the second and third layers help prevent an attacker from being successful. This makes it more difficult for someone to penetrate a network because there are layers of security built into every system, process and piece of the network architecture. AT&T utilises this "defence in depth" architecture with proactive management to quickly and easily determine if known threats are being directed at our network or may affect application performance and then mitigate these threats before they can affect the network or the applications running across it. This way we can help reduce the risks to our network and help protect our customers' network connections and the information crossing it.

A Strategy for Enterprise Communications Is Critical for Deeper Broadband Deployment

A summary of the key messages delivered in this chapter are as follows:

1. Enterprises have a role to play: by pushing the boundary on productivity initiatives that collectively constitute demand that makes infrastructure providers take notice.

2. Enterprises require a supportive eco-system: Infrastructure development and policy all have a role to play in providing an environment conducive to advancing growth and social benefit.
3. Cooperation and partnerships are critical: No single organisation can lead such a radical initiative; it will require collaboration amongst organisations with complementary strengths as well as governmental support.

Chapter 9
The Impact of Policy on the Performance of the ICT Sector

Raúl L. Katz

Introduction

This chapter discusses and provides evidence surrounding the link between public policy initiatives, ICT sector performance, and its corresponding economic impact. In particular, it investigates the way in which policies might influence ICT diffusion and adoption, examining models of regulation and public policy and their relationship to specific sector performance. In doing so, it identifies certain public policies and frameworks consistently associated with above par sector performance. Going beyond specific policies, it aims to understand why some countries are more effective than others in implementing policy tools.

The evidence generated in this paper is critical in terms of the creation of the right conditions for developing next generation broadband to serve as a video platform in emerging regions such as Africa. Data presented in this paper will support the notion that public policy, if implemented appropriately both from an institutional and specific initiative perspectives, has the potential for leading emerging countries to leapfrog in technology development.

Relying on both descriptive statistics and case studies, it recognizes that some countries have had more success in implementing policy tools than others but points to key commonalities amongst the nations with the strongest performing ICT sectors. While the statistical analysis employed provides quantitative evidence of these links, the case study data uncover the institutional and cultural variables that can impact ICT sector growth, many of which may go unnoticed in purely quantitative analysis. The chapter also looks at trends in policy evolution, discussing the

R.L. Katz (✉)
Adjunct Professor, Division of Finance and Economics, Columbia Business School, 3022 Broadway, New York, NY 10027, USA

Director, Business Strategy Research, Columbia Institute for Tele-Information, 3022 Broadway, New York, NY 10027, USA
e-mail: rk2337@columbia.edu

J. O'Neill et al. (eds.), *Broadband as a Video Platform*, The Economics of Information, Communication, and Entertainment, DOI 10.1007/978-3-319-03617-5_9,

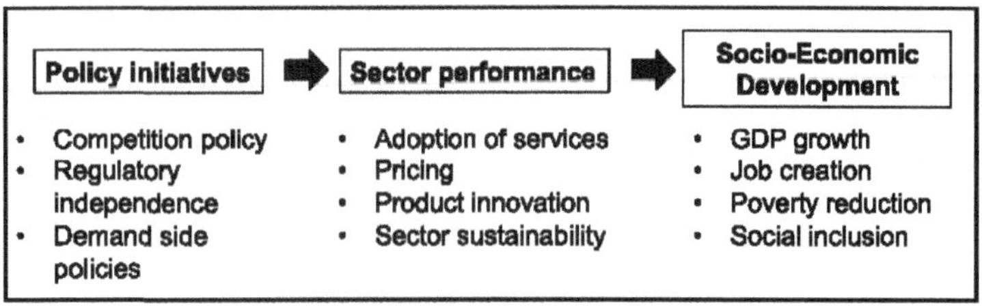

Fig. 9.1 Causal chain between ICT public policy and economic impact

importance of government planning, sector vision, and a clear blueprint as well as the need for accountability and leadership.

Ultimately, it concludes that a country must adopt certain best practices in order to maximize the impact that policy will have on ICT sector performance, which will then lead to economic growth. In other words, while public policy and regulation drive the performance of the ICT sector, active government involvement must complement these policies.

The chapter provides evidence regarding the linkage existing between public policy, ICT sector performance, and economic impact along the following causal chain (see Fig. 9.1).

It proceeds from providing evidence of the impact of ICT sector performance on the economy, focusing after on the influence public policy has on the performance of the sector.

Examining the Link Between ICT Sector Performance and Economic Development

The contribution of ICT to the economy has been studied for some time.[1] In recent years, evidence in this domain has also been generated for African countries. Our own 2012 study of wireless access and the economy in Senegal,[2] for instance, demonstrates the effects of sector performance on economic growth and compiling country-level time series on technology adoption and economic performance from 2004 to 2011. The study suggests that mobile phones have a measurable impact on economic growth and lie within the estimates of previous work on a much larger scale. The study determined that the Senegalese GDP grew approximately 0.044 % for every 1 % increase in the country's mobile penetration. On the other hand, the economic effect of broadband could not be measured yet because the technology

[1] See Katz (2012); *The Impact of Broadband on the Economy: Research to date and Policy Issues*. Geneva: International Telecommunication Union for a review of the literature.

[2] Katz, R. and Koutroumpis, P. (2012). The economic impact of communications in Senegal. *Communications and Strategies*.

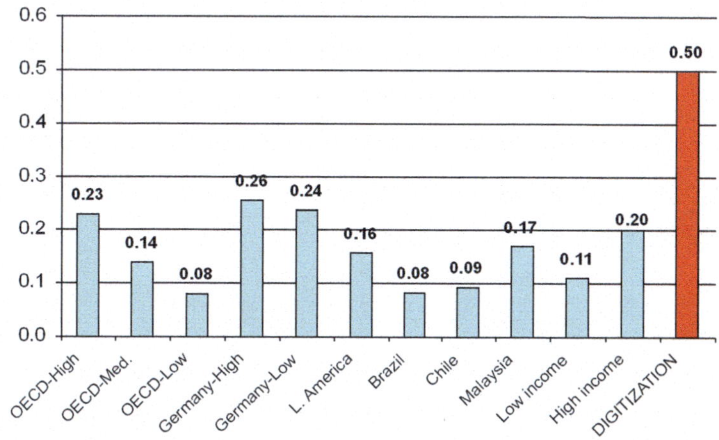

Fig. 9.2 Digitization and economic development (contribution to GDP growth of 10-point increase in variable)

was at its very early stages of adoption. However, the rapid growth of third-generation services during 2011 would suggest a transformation in this type of network access, which might result in important economic effects in the future.[3]

In recent years, the study of the economic impact of ICT has shifted toward an understanding of the contribution of digitization and away from discrete technologies such as wireless or broadband.[4] Digitization refers to the cumulative adoption of all information and communication technologies in addition to the assimilation and usage in the economic and social fabric. Figure 9.2 provides a comparison of the magnitude of impact of GDP per capita growth across discrete telecommunication technologies and digitization.

Examining studies that cover both broadband and mobile studies in multiple countries, we found that a 10-point increase in digitization yields a 0.5 % increase in GDP per capita, a magnitude significantly larger than that of either broadband or wireless. This result signifies that only through digitization, which incorporates applications and services, is the full impact of ICT realized.

Digitization also impacts employment, with full deployment and assimilation of ICT contributing to sector jobs such as software development, business process outsourcing, equipment manufacturing, and parts supplies. As is the case with GDP growth, digitization has a higher impact on job creation than broadband deployment alone. On average, a 10-point increase in digitization results in a 0.82 % decrease in a country's unemployment (Fig. 9.3).

---

[3] See new evidence in Chap. 11 of this volume.

[4] Katz, R., Koutroumpis, P. and Callorda, F. (2013). The Latin American path to digitization. *Info*. Vol. 15 No. 3, pp. 6–24; Katz, R., and Koutroumpis, P. Measuring socio-economic digitization: a paradigm shift, *Technovation* (in process).

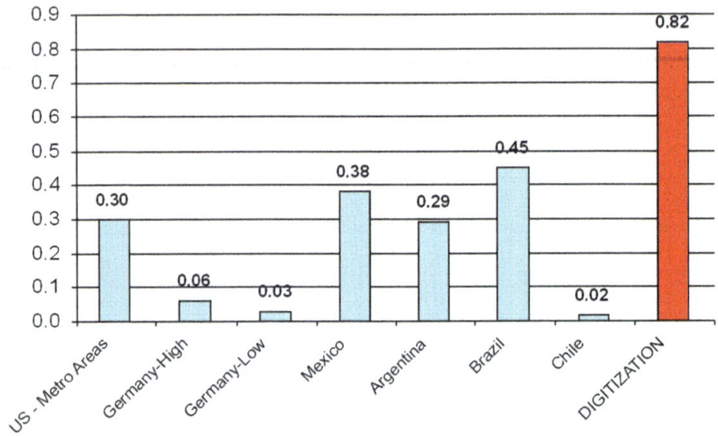

Fig. 9.3 Digitization and employment (contribution to job creation of 10-point increase in variable)

At this point, it is therefore safe to conclude that the impact of ICT, whether at the discrete technology platform level (such as wireless or broadband) or at the aggregate level of digitization, is quite significant. The issue that needs to be raised now is if public policy constitutes a lever through which the aforementioned economic impact can be maximized.

Examining the Link Between Policy and ICT Performance

If ICT has an impact on the economy, can it be proven that public policy has an influence on the performance of the ICT sector? In other words, if the extent of ICT economic contribution is driven by the performance of the ICT sector (in terms of lower prices, enhanced adoption of technology, speed of innovation, and sustainability of the sector), can public policy affect performance? For this purpose, the first task was measuring the sector's performance level. This objective was achieved by developing an index comprising multiple indicators, grouped into three large clusters:

- ICT adoption: comprising broadband and wireless penetration, broadband prices (advertised and effective), and wireless prices (service revenue per minute).
- ICT innovation including percent of mobile average revenue per user (ARPU) derived from data services and quality of service metrics (mobile: dropped calls, service coverage; wireline: Average speed of answer in customer care calls, mean time to repair).
- ICT sector economic performance comprising average wireless sector EBITDA, incumbent investment in broadband and next-generation networks, entrants' investment in broadband and next-generation networks, and mobile carriers' capital investment.

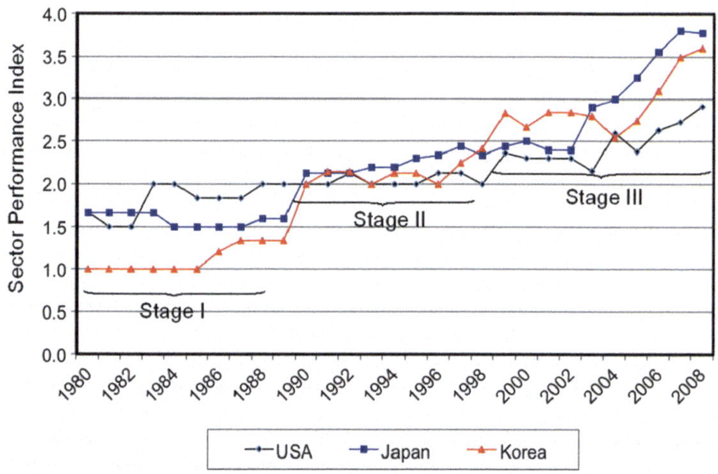

Fig. 9.4 Evolution of sector performance: The United States, Japan, Korea (1980–2008, plotted against the Sector Performance Index)

The index was used to examine the sector performance of 52 countries in the past 30 years. In particular, it was attempted to find answers to questions such as the following:

- How have countries like Korea and Japan performed relative to other industrialized nations? What policy and regulatory variables could explain their different relative performance?
- Are the "BRICs"[5] behaving homogeneously? Is there a consistent or a divergent development path? Are policies affecting performance?
- What is the path toward enhanced sector performance of emerging countries like? Is there a uniform development path? What are the consistent patterns?

In all, the examination of the evolution of the ICT performance index over time yielded several interesting observations. First, as Fig. 9.4 indicates, the Japanese and Korean ICT sectors bypassed that of the United States in 1998. A comparison of the index over time for the three countries reveals three development stages. Between 1980 and 1990, the performance of the ICT sector of the United States surpassed that of Japan and Korea. After 1990 and approximately up to the year 2000, all three countries exhibited comparable performance. Finally, after 2000, Korea and Japan surpassed the United States. We will explore later what variables can explain this reversal of fortune.

The next analysis compares the performance of the countries known as BRIC (see Fig. 9.5). The most startling feature of this comparison is how China, after consistently being a laggard within this group until 1992, surpassed the other countries in terms of ICT sector performance in 2004.

[5] BRICs – Brazil, Russia, India, and China.

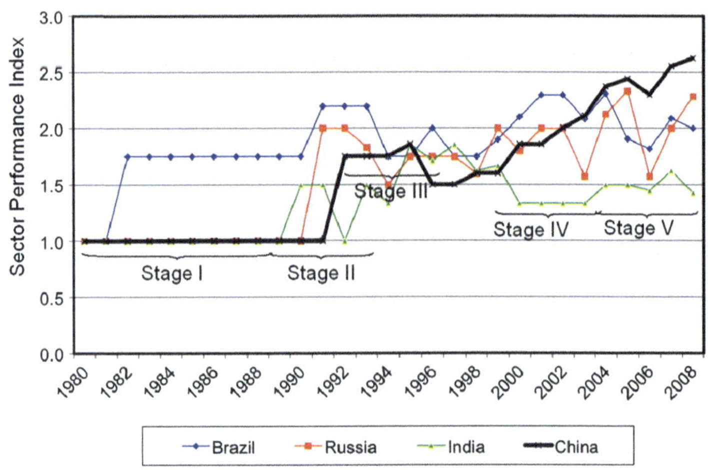

Fig. 9.5 Evolution of ICT sector performance: China versus BRICs (1980–2008, plotted against the Sector Performance Index)

Can we relate these two processes, the Japanese and Korean as well as the Chinese, to specific policy variables? For this analysis, we need to understand what was occurring in the policy arena in each of these countries.

Understanding ICT Policy Evolution

Multiple causal links clearly exist between ICT policy and sector performance. This study would not be the first to find such links. Research has shown, for example, that the institutional framework impacts technology adoption as regulatory autonomy appears to reduce prices and improve wireless penetration and privatization improves wireline penetration. It also impacts the innovation and economic clusters, as an independent regulatory agency and privatization tend to improve sector economic performance, particularly investment.[6]

[6] See for example, Boyle, G., Howell, B., and Zhang, W. (2008). *Catching Up in Broadband Regressions: Does Local Loop Unbundling Really Lead to Material Increases in OECD Broadband Uptake?* NZ Institute for the Study of Competition and Regulation; Cava-Ferreruela I., Alabau-Munoz, A. (2006). "Broadband policy assessment: A cross-national empirical analysis" *Telecommunications Policy* 30:445–463; Distaso, W., Upi, P., and Manenti, F. (2006). "Platform competition and broadband uptake: Theory and empirical evidence from the European Union" *Information Economics and Policy* 18:87–106; Garcia-Murillo, M.(2005). "International Broadband Deployment: The Impact of Unbundling" *Communications & Strategies,* No 57 1st quarter; Grzybowski, L. (2005). "Regulation of Mobile Telephony across the European Union: An Empirical Analysis" *Journal of Regulatory Economics;* 28:147–167; Grzybowski, L. (2008). "The impact of regulation on the retail prices in fixed-line telephony across the European Union", *Telecommunications Policy* 32:131–144; Gutierrez, L. (2003). "The Effect of Endogenous Regulation on Telecommunications Expansion and Efficiency in Latin America." *Journal of*

The regulatory framework plays a role as well, as competition can increase wireless penetration, number portability can increase wireless and wireline prices, and platform competition can increase broadband adoption. Furthermore, policies that encourage competition impact wireline and wireless deployment, while access regulation discourages investment.

That said, neither the impact of policy on innovation nor the holistic impact of policy on sector performance has undergone in-depth analysis. In terms of the policy variables, the institutional and regulatory framework could influence innovation, but there is not, to our knowledge, an assessment of the impact of regulation and policy variables on the rate of ICT innovation. Non-sector-specific policies could also shape all three of the ICT sector performance metrics, but studies need to examine the impact of trade regulation and performance and the impact on convergence regulation and performance. Thus, a comprehensive study of the relationship between all regulatory and policy variables and full sector performance would contribute to the field of existing research.

To assess country performance in the ICT policy domain, an index was developed, integrating three components:

- The institutional framework: This component includes such factors as the type of governmental entities that are in charge of developing ICT policy or regulating the sector and the providers of service. For example, is there a cabinet-level position centralizing all ICT policy matters? What is the scope of the telecommunications national regulatory authority and its enforcement powers? Is there an overarching ICT national planning process?
- The ICT regulatory framework: This component comprises all of the variables related to specific ICT policies and regulatory approaches. They include regulations related to market entry (e.g., vertical separation, local loop unbundling, rights of way, numbering schemes, spectrum management), price regulation (interconnection, mobile termination rates, weighted average cost of capital, retail pricing), investment incentive regulation (e.g., asymmetry), actual regulatory process (e.g., market analysis ex ante), and methods of applying regulations (e.g., technological neutrality, operational conditions, compliance monitoring).
- Policies not specific to the sector but having some spillover effect: The policies in this component include restrictions on foreign direct investment that affect market entry and capital structure and other trade restrictions that affect services supply and long-term government planning, as well as the regulation of audiovisual content that impinges on convergence (e.g., restrictions on telecommunications carriers regarding content distribution). In addition, policies that promote and facilitate the adoption of ICT by late adopters (the poor and small and medium enterprises), such as digital literacy programs and equipment subsidization, can play an extremely important role in fostering the emergence of a high-performance sector.

Regulatory Economics, 23:32 57–3286; Katz, R. (2009). *El papel de las TIC en el desarrollo*. Barcelona: Ariel; Li, W. and Xu, L. (2004). "The impact of privatization and competition in the telecommunications sector around the world." *Journal of Law and Economics*, 47, 395–430.

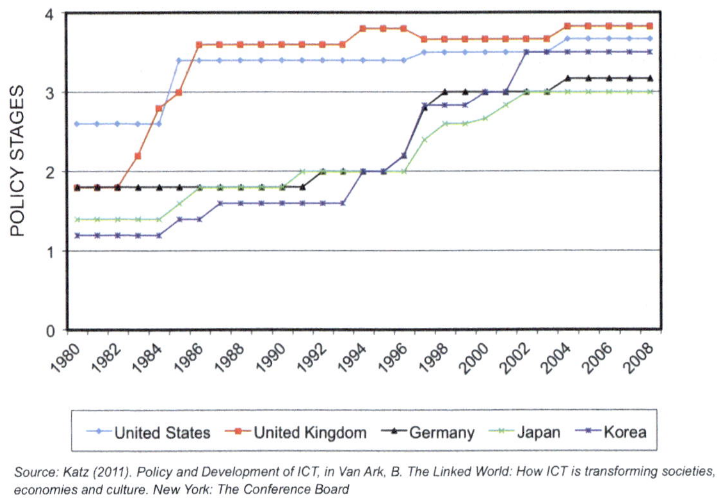

Fig. 9.6 Evolution of Policy Index (1980–2008)

Figure 9.6 presents the evolution of the policy index for five countries between 1980 and 2008. According to the time series of the index in Fig. 9.6, countries appear to follow distinct ICT policy evolution paths. For example, in countries following the so-called *Anglo-Saxon path*, like the United States and the United Kingdom, the policy index jumped as a result of early liberalization and privatization. These countries privatized the incumbent and opened the markets to competition early on while employing competitive safeguards. In 1984, these two countries pushed for market opening in an effort that culminated in the 1990s and resulted in near-full liberalization.

In contrast, countries following the *gradual liberalization path*, like Germany and Japan, saw both liberalization and protection of the incumbent along with a step-by-step adoption of deregulation and gradual privatization. Until the mid-1990s, the telecommunications sectors of these two countries were served by state-owned monopolies. At this point in time, the state-owned telecom providers were immune to competition and maintained high investments and staffing levels to provide the necessary advanced infrastructure and services. In the mid-1990s, Japan and Germany embarked on a process of rapid liberalization, with regulators focusing less on protecting the incumbent and more on establishing competition.

Finally, countries following the *developing path*, like Korea, gradually liberalized the telecommunications sector in a methodical fashion to allow for planned performance improvements.

Is it possible to establish a relationship between the policy index and ICT sector performance? When controlling for economic development, a direct relationship seems to exist between both indices. In other words, the higher the policy index along institutional, regulatory, and non-sector policy impacts, the higher the

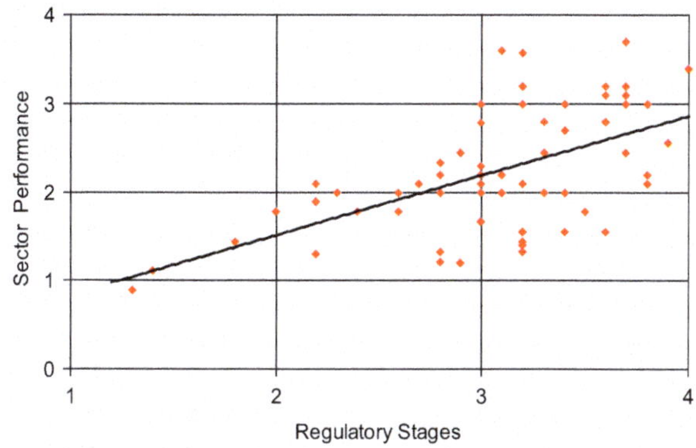

Fig. 9.7 Relation between policy stages and ICT sector performance levels (2008)

performance of the ICT sector in terms of lower prices, technology adoption, and degree of innovation (see Fig. 9.7).

Figure 9.7 demonstrates the distribution of the countries studied, with the countries with the least advanced policy levels occupying the space reserved for countries with the lowest ICT sector performance. On the other hand, those countries with the highest sector performance also had the most advanced policy index scores. These countries included Australia, Austria, Denmark, Finland, France, Germany, Italy, Japan, Korea, Portugal, Sweden, the United Kingdom, and the United States.

The policies of these countries tended to exhibit the following commonalities:

- Full competition in all telecommunications industry segments
- Universal service obligations for fixed line and broadband, driven by a fair allocation of contribution across industry players
- A privatized telecommunications incumbent
- Voice over IP telephony allowed with regulation in place
- No restrictions on foreign ownership of industry players
- A comprehensive national plan to promote ICT industries (software, services, applications)

Beyond this static relationship, case study data also highlights the direct influence of policy on sector performance. The case of China is particularly noteworthy, given that, as mentioned above, in just 15 years, this country managed to advance from a laggard amongst its emerging country counterparts to take the lead of this group and become the top performer amongst the BRICs. The development of the ICT sector has been strongly linked to China's strong top-level institutional leadership. The Ministry of Industry and Information Technology took responsibility for the development of ICT policy, reporting in with the State Council. It is also a member of the State Informatization Leading Group (SILG), which approves regulatory decisions and oversees the industry, playing a particularly significant role in the Internet regulation and the control of web content.[7,8]

Taking into account the differences in the culture, political structure and regulatory advancement of different countries between African countries and China, the Chinese example is relevant insofar that it depicts how a country that is lagging in the development of the ICT sector can rapidly achieve a leadership position. The next chapter will explore what practices identified in this study could be adopted by African countries to build next generation broadband network platforms.

Policy Implications for Building a High-Performance ICT Sector

If policy can help maximize the performance of the ICT sector, and, therefore, its economic contribution, it is pertinent to examine which policies and best practices in the policy domain can be more fruitful. Case study data enabled identification of several prescriptive implications.

Firstly, in general, policies (1) promoting the adoption of competition, (2) guaranteed by regulatory independence, and (3) guided by an overarching vision for the ICT sector appear to be linked to ICT sector performance. The countries with the highest ICT sector performance levels tend to have the following commonalities in terms of policy and regulatory features:

- Full competition in all telecommunications industry segments
- Universal service obligations for fixed line and broadband, driven by a fair allocation of contribution across industry players
- A privatized telecommunications incumbent
- Voice over IP telephony allowed with regulation in place
- No restrictions on foreign ownership of industry players
- A comprehensive national plan to promote ICT industries (software, services, applications)

That said, not all telecommunications competition models see equal success in stimulating investment and innovation. While competition encourages investment, at some point, too high a level of competitive intensity decreases the incentive to invest and deploy wireless broadband services. A certain amount of market concentration and a moderate level of competitive intensity appear to be most effective in driving deployment. The implication of this approach to competition policy in Africa is that governments should not be concerned with promoting unrestricted competitive market structures. On the contrary, a moderately concentrated industry

[7] Source: Katz (2011). Policy and Development of ICT, in Van Ark, B (editor). *The Linked World: How ICT is transforming societies, economies and culture*. New York: The Conference Board.

[8] See Katz, R., Hoffman, D., Jaeger, K. *China: an ICT "catch up" strategy through a state-owned sector and centralized planning*. Case study in support of The Linked World (op.cit.).

could generate the necessary static and dynamic benefits to consumers, while ensuring sustainability to market players.

Secondly, *competition policy and regulation cannot reach their full potential without active government involvement.* Government planning and an explicit target or vision can play a critical role in driving sector performance. The Korean Government, for instance, has prepared 5-year plans since 1995 to address such issues as universal broadband and its position in the global IT market. Similarly, Japan launched its e-Japan Strategy in 2001, developing annual plans to address the same concerns. African countries typically lack governments actively involved in the formulation of multi-year state policies in the area of ICT. The development of next generation broadband platforms requires that governments, in combination with consumers and the private sector, get involved in outlining a long-term vision of technology development, supported by plans and targets.

Beyond proactively planning for the future, countries must continue to follow up and assess these plans once implemented. At the end of the forecasted period for each plan, the Korean Government assesses the results and then incorporates this feedback into future models. In China, government-sponsored planning supports institutional centralization. Senior leadership reviews are based on the achievement of the plans' quantifiable goals, relating to such indicators as network capacity expansion, coverage, and penetration as well as quality standards. Meanwhile, to ensure sector success, some governments in countries such as Korea, China, and Japan take their intervention to the next level, actively shaping the industry structure, through directives on mergers and acquisitions. Again, African governments tend to shy away from proactive planning and disciplined monitoring of implementation. If plans are to be effective, they need to be followed by an ongoing monitoring of their fulfillment.

Thirdly, *governments can play a key role in ICT adoption through the implementation of demand-side policies.* By identifying government entities that require broadband services (e.g., administration, public schools, hospitals) and turning them into anchor tenants, for instance, they can ensure that broadband investment can reach a breakeven point, as seen in Korea and the Netherlands. Similarly, Sweden and Japan offered tax incentives to companies in the ICT assets and software business to stimulate investment in the sector.

The development of e-government services has also encouraged ICT adoption and influenced the demand side of the equation. Korea, Estonia, and Colombia have demonstrated notable commitment to connecting their citizens. Some examples of these services include tax return e-submission, an e-procurement service for SMEs selling goods and services to the government, platforms for telecommuting, and platforms that support e-business transactions between the government and enterprises. Looking specifically at SMEs, Japan encouraged these businesses to implement IT services voluntarily by offering training, promoting best practices, and supporting collaboration with local communities.

Targeting specific sectors of the population also leads to an increased demand for ICT services. Specifically, offering digital literacy programs, computer subsidies, and online education programs can spur adoption amongst citizens otherwise hindered by socioeconomic and cultural limitations, such as the elderly or the disabled. Sweden, Estonia, and Korea—all of which have high rates of ICT adoption—have all actively promoted demand-side policies that emphasize digital literacy, subsidization of access, and promotion of applications that stimulate adoption. With a few notable exceptions, such as the stimuli deployed by Kenya in the development of applications new ventures, African countries lack a concerted effort to foster the incubation of start-ups.

Lastly, *executive branch leadership and clear articulation of regulatory and industrial policies* can enhance a country's ICT sector performance. For instance, solid industrial policies can link the development of a telecom sector and the creation of export-oriented IT services and software industries. Korea funds the adoption of its products in the domestic market, thereby incubating its export-oriented industry. In 2007, the MIC in Japan established the ICT International Competiveness Enhancement Program, collaborating with industry, academic, and government leaders to promote its products in developing markets. To support its domestic technology cluster, the Estonian Government sponsored the Competence Centre in Electronics-, Info-, and Communication Technologies (ELIKO) in 2004. Chinese policy makers focused on cultivating state-owned telecom equipment champions by consolidating its Ministry of Electronic Industries (MEI) to create the Ministry of Information Industries (MII).

Support of ICT policy from the executive branch also appears to play a role in high-performance sectors. This top-level leadership can give the sector direction and ensure cooperation and consolidation amongst all branches of the government, aiding in the achievement of national digital goals. Korea, for instance, names an "ICT Czar." This official regularly meets with the Korean President, who in turn takes ultimate responsibility for sector development. In Brazil, the Secretariat for Strategic Affairs of the Presidency of the Republic and the Casa Civil partnered to create the country's National Broadband Plan, which the president approves directly. Chinese ICT sector development has been attributed to strong executive leadership. Here, the Communist Party controls all senior management personnel decisions in order to ensure compliance. It is fairly common in African countries to delegate responsibility for fulfillment of ICT plans to lower level government entities. This approach reduces significantly the impact of a planning activity or policy implementation.

All things considered, our research has demonstrated the importance of public policy in developing a strong, high-performance ICT sector. It recognizes, however, that different nations exhibit different policies and practices, many of which are influenced by their respective political systems. At the same time, despite these differences, certain best practices seem to occur across the most successful countries. These practices are crucial in maximizing the impact of policy on performance, which ultimately leads to a subsequent economic contribution. Best practices, thus, support performance leapfrogging and maximize the economic impact of ICT development.

If one looks at the higher level of analysis, digitization brings diversity that allows countries to enhance all sectors of public welfare as well as economic growth using ICT, such as using the internet on fixed as well as mobile platforms as a platform for video and all the sectors broadband video can serve. The implication of these findings for the development of next generation broadband as a video platform in Africa are self-evident.

Chapter 10
Information Technologies for Rural Development in Africa: New Paradigms from Radio to Broadband

Heather E. Hudson

Introduction

Developing regions face many critical challenges, of which the most basic is producing sufficient food for their people. Some 240 million people in sub-Saharan Africa don't have enough to eat; three-quarters of them live in rural areas. They depend on farming to feed themselves and to generate income for other needs on land that is often prone to natural disasters such as drought or floods. The UN's Food and Agriculture Organization (FAO) states that Africa is in a food security crisis (Food and Agriculture Organization 2012).

Many strategies have been developed that could address food security, such as selecting crops that provide more nutrition, using seeds and techniques that resist drought, using local products as fertilizer, and using improved methods of processing and storage. In addition to improving crop yield, farmers may need new strategies to generate income from their crops and livestock. Improvement in food security therefore requires transmission of information, so that farmers are aware of these new practices and techniques. Yet awareness alone is not sufficient; farmers must actually adopt these new practices.

Radio remains the most widely used medium in rural Africa, reaching people without electricity and those who are illiterate. In the communities participating in the project described below, approximately 76 % of households owned a radio (Farm Radio International 2011a). Radio therefore seemed an appropriate choice

Research reported in this chapter was carried out by Farm Radio International (FRI) with support from the Bill & Melinda Gates Foundation. The author is a member of the board of FRI and was an advisor on the project evaluation.

H.E. Hudson (✉)
Institute of Social and Economic Research, University of Alaska Anchorage, Anchorage, AK 99508, USA
e-mail: heatherehudson@gmail.com; hehudson@uaa.alaska.edu

J. O'Neill et al. (eds.), *Broadband as a Video Platform*, The Economics of Information, Communication, and Entertainment, DOI 10.1007/978-3-319-03617-5_10,

for creating awareness of best practices to enhance food security among African farmers.

Radio for Development and ICT4D

Electronic mass media have been used as a means of disseminating development information since the early days of radio broadcasting. Radio programs, often in conjunction with agricultural extension and other outreach activities, have been frequent components of agricultural campaigns. Radio broadcasts were included in development campaigns and as components of distance education curricula. Interactivity was found to increase learning and adoption of new practices through such techniques as listening groups, call-in programs, and classroom activities to accompany instructional radio programs. [See, for example, the writings of Schramm (1964) and the recent synthesis on media and development by McAnany (2012).]

In the past decade, mass media have been joined by other information and communication technologies (ICTs) that have been used for development initiatives and are referred to as ICT4D. [See a summary of this research by Hudson (2006).] Development agencies such as the FAO drew attention to the potential of new applications such as radio streaming and websites and other materials available online (Girard et al. 2003). In 2008, the Panos Institute of West Africa (PIWA) carried out a survey of 220 radio stations in seven West African countries concerning ICT use and Internet connectivity. The study noted the need to strengthen human capacity to use these new technologies for development: "The limitations are often due to the lack of awareness of the possibilities offered by ICTs, as well as a skill gap in the staff able to deliver expected services" (Ndiaye et al. 2008).

Examples of recent ICT innovations to support radio in developing regions include Gramin Inter-Networking System (GRINS), a software suite developed in India for community radio integrating mobile phone technology, and Freedom Fone, an open-source interactive voice response (IVR) system developed in Zimbabwe (Farm Radio International 2011a). The latter was included in the pilot projects described below.

The African Farm Radio Research Initiative

Could radio also help to increase adoption of agricultural practices intended to improve food security in Africa? And what about newer technologies that might enhance radio's effectiveness, such as mobile phones, portable digital audio recorders, and Internet access at radio stations? Farm Radio International (FRI), a Canada-based NGO which had provided training and agricultural program content for African radio stations for several decades, undertook a 4-year project called the

African Farm Radio Research Initiative (AFRRI) to answer these questions, with support from the Bill & Melinda Gates Foundation. Specifically, AFRRI addressed the following:

1. How effective is radio in enabling smallholder farmers in Africa to address food security challenges they face, with a particular focus on increasing/diversifying food production, improving land use management, and reducing post-harvest losses?
2. How can new technologies, such as cell phones and MP3 players, increase the effectiveness of radio as a sustainable, interactive development communications tool? (Farm Radio International 2011b)

AFRRI partnered with 25 radio stations reaching an estimated 40 million farmers in five countries: Ghana, Malawi, Tanzania, Uganda, and Mali. One of the key elements of the project was participatory radio campaigns (PRCs) designed to involve farmers in every aspect of the planning and production of agricultural radio programs. A second was a radio-based information service to help farmers obtain current and relevant information on markets and prices for their produce. A third key element was ICT-enhanced farm radio, which involved testing various equipment packages at the radio stations.

The research methodology for the PRCs involved identifying three clusters of communities for each of the 25 radio stations. Active listening communities (ALCs) included farmers who were interviewed about their agricultural practices and needs and their radio listening habits. They then were invited to be involved in the design of a series of radio programs addressing a particular agricultural practice to help improve their livelihoods and ultimately their food security. A second cluster was designated passive listening communities (PLCs), where farmers listened to the programs but did not participate in planning or any interactive follow-up. A third cluster consisted of control communities that did not have access to the radio programs. Follow-up surveys after each of the two 16-week campaigns were designed to collect data on the extent to which farmers learned about the agricultural innovations and actually adopted the new agricultural practices.

Farmers engaged in the design and development of farm radio programming were almost 50 % more likely to take up agricultural practices deemed to improve their food security than passive listeners. Those in the ALCs were ten times more likely to adopt the practice than those farmers who had no access to the farm radio programs (see Chart 10.1).

Lessons from the AFRRI's Participatory Radio Campaigns

FRI defined a PRC as "a planned, radio-based activity, conducted over a specific period of time, in which a broad population of farmers is encouraged to make an informed decision about adopting a specific improvement selected by their peers, based upon the best available information, to improve the food security of their

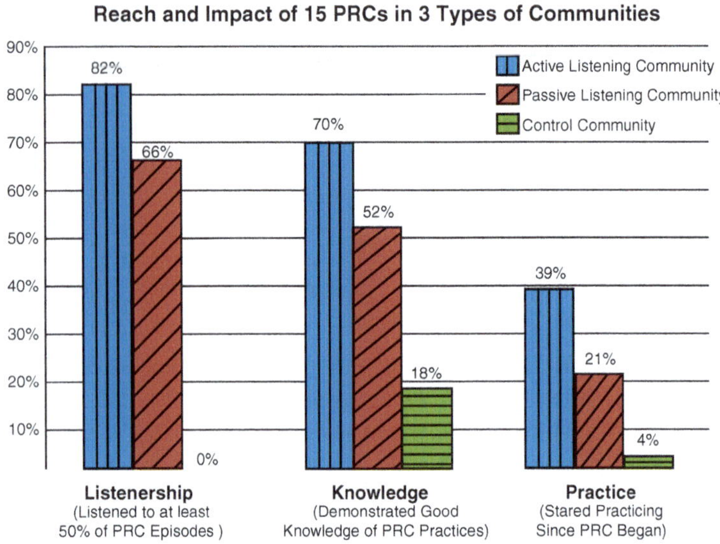

Chart 10.1 Comparative impact in three types of communities. *Source*: Farm Radio International (2011b)

families. It then provides the adopting farmers with the information and other support they require to implement the improvement" (Ward 2010). This definition is not necessarily limited to farming; it could obviously be broadened to apply to a strategy to foster adoption of any practice by a specific target population.

Similarly, the lessons from the AFRRI's PRC activities could also be generalized and adapted to other contexts. Among the best practices identified from the first set of campaigns were the following:

- *Using the farmers' language*: More generically, it is important to use the language spoken and understood by the target audience.
- *Role of audience participation*: Consulting representatives of the target audience about the campaign and the programming can increase participation in the campaign.
- *Importance of audience voices*: Listeners like to hear their voices and voices of others facing similar problems. There are many ways of capturing these voices, through taped interviews, call-in and callout programs, and other techniques discussed in the ICT section below. As AFRRI points out: "These voices provide credibility and attractiveness and encourage farmers to engage in the campaign … They reveal the struggle each farmer undergoes as s/he works towards a decision about implementing the improvement. And these voices of farmers explain how they are implementing the improvement, and overcoming problems that crop up while they implement" (Ward 2010).
- *Role of expert information and advice*: Target audiences also need to hear from experts with knowledge relevant to the campaign. Experts can be interviewed in

the studio or in the field with a target group or wherever they may be using mobile phones. To be effective, these experts must have good presentation skills and use language and examples familiar to their audience in order not to bore or confuse the audience.

- *Role of entertainment*: Effective programs must be entertaining; in addition to relevant and well-presented information, they can include local music, stories, drama, humor, or other entertaining content.
- *Regularity and repetition*: Programs should be broadcast according to a regular schedule at a time when most of the audience can listen. Repetition is also important to catch audience members otherwise unavailable. In addition to repeat broadcasts, making programs available on portable audio players, or where facilities exist, through podcasts and audio streaming can increase audience access. Other techniques and media such as flyers, songs, and discussion groups can help to repeat and reinforce the campaign messages (Farm Radio International 2011b).

ICT-Enhanced Radio Experiments

Another major component of AFRRI was a series of experiments to determine whether a combination of radio plus other ICTs could enhance the effectiveness of the PRCs. Radio is a one-way medium, whereas participation, as noted in the results of the PRC project above, can increase both awareness and adoption of new agricultural practices. Mobile phones are becoming increasingly available in Africa, including rural areas, and provide a means of interacting with the radio audience such as for call-in questions, callout interviews, and text messages to remind farmers about program schedules. Could they be part of a participation strategy? Radio stations in Africa typically have not had access to integrated mobile phone technology that would facilitate these applications; many may also not be able to afford significant mobile phone charges.

Also portable digital recorders (MP3 players/recorders) could be used not only for interviews with farmers but also for feedback from them on the radio programs. Digital players could also provide a form of radio-on-demand for farmers who missed the scheduled broadcasts. Software that runs on personal computers can be used for digital editing. However, as FRI notes: "There are still many radio stations in Africa that rely on tape recorders, large expensive batteries, and often broadcast direct-to-air. They lack editing equipment, and the skills to produce and prepare shows in advance of a broadcast. These processes limit the ability of radio stations to produce good farm radio programs for their listeners." Internet connectivity could provide the radio stations with access to agricultural content, training materials, and inexpensive interaction with experts and colleagues through e-mail and Skype.

The purpose of the ICT experiments was to test how new ICTs could be integrated with radio to provide better interactive communication between the radio stations and their target farmer audience. AFRRI provided eight customized

ICT packages in experiments with the 25 radio stations in the project. The following were included in the various ICT packages:

- *Computer and Internet access*: Each partner radio station received a desktop computer and Internet access. Participating broadcasters received training in basic computer literacy, Internet search skills, and virus protection techniques. Some 68 % of 51 partner broadcasters surveyed identified the Internet as the most important ICT tool for production of farm radio programs (Farm Radio International 2011a).
- *Digital recording and editing equipment*: Rechargeable MP3 recorders were provided to all broadcasters participating in AFRRI. They were very popular with the broadcasters, enabling them to capture high-quality audio such as interviews with farmers and extension agents, which could be edited into the radio programs. FRI notes: "The portability of MP3s allowed radio hosts to visit farmers in their fields and in their homes, increasing farmer participation in the radio broadcasts" (Farm Radio International 2011a).
- *Phone callouts to extension agents and experts*: Although many radio stations had started to include call-in comments from listeners with mobile phones, they had not used phones themselves to reach pre-identified resource people for interviews and commentary. Using mobile phones enabled extension agents to participate in the programs without having to travel to the stations, which was both time consuming for the agents and often expensive for the stations if they had to pay travel expenses. Of 41 extension agents surveyed, 61 % believed that the reach and impact of their extension work were substantially improved because they could be heard on radio programs through callout programs.
- *Phone callouts to farmers*: Callouts to farmers with mobile phones enabled farmers to learn from other farmers. Typically, broadcasters made prearranged calls to two or three farmers per weekly episode. AFRRI evaluation found that callouts to farmers can significantly affect farming adoption rate of agricultural improvements (up to 14 %), increase their level of knowledge about the agricultural practice (up to 50 %), and improve their overall listenership (up to 22 %), when compared with stations that do not make callouts to farmers. There were also significant cost savings compared with traveling to villages; AFRRI found that the cost of calling three farmers for 5 minutes each was approximately $US8 compared to $US75 to travel to three villages for interviews (Farm Radio International 2011a).
- *SMS alerts for farmers*: AFRRI provided access to services that enabled broadcasters to send SMS alerts to listeners' mobile phones 30 minutes before the program. Recipients were asked to share the reminder with their neighbors. AFRRI found that these weekly SMS alerts were a cost-effective means of increasing listenership by up to 20 %; in turn, listening to more episodes was correlated with higher levels of adoption of the practices in the programs. Messages cost about $US0.05 each, so that reminding a farmer who in turn shared the information with neighbors cost less than $US1 for an 18-week campaign. However, several steps are required to implement this procedure,

such as compiling a database of phone numbers of farmers willing to participate and accessing a service that would automate the mass SMS calls [(for details on equipment and service, see Farm Radio International (2011a))].

- *Local agents with solar-powered radios/MP3 recorders and mobile phones*: Some communities nominated a local resident to be a radio agent to host listening sessions during campaign broadcasts, record programs for repeat listening at a later time, and provide access to a mobile phone for communication with the radio station. (This model of community listening groups has a long history in radio for rural development.) Radio agents were typically women; PRC research had found that in some communities men took family radios to the fields, thus depriving women of the opportunity to hear the programs. A radio with recording capacity enhanced the value for the listeners, as programs could also be recorded for later or repeat listening. Of farmers surveyed, 73 % reported that group listening helped them better understand the content through discussion with the group, while 46 % of the same group of farmers stated that listening in a group gave them encouragement to start practicing the agricultural improvement (Farm Radio International 2011a).
- *IVR*: Two radio stations, one in Ghana and the other in Tanzania, experimented with using interactive voice response, enabling farmers to call in from mobile phones and access agricultural information on demand through a series of menus. This experiment indicated that some farmers were willing to use mobile airtime to access agricultural information on demand or leave messages about content, but users tended to be younger males with secondary education, indicating that this type of ICT application was not as readily adopted as voice calling and text messaging. Also, some respondents including 35 % of women said that they could not afford the calls.
- *Connectivity*: *Satellite terminals and fixed wireless*: For some radio stations access to the Internet via satellite was the only option. AFRRI experimented with technology that could enable such stations to sustain the costs for satellite access. Two radio stations, one in Tanzania and the other in Mali, were equipped with VSATs (small satellite terminals) and fixed wireless to extend Internet access. The stations set up and sold wireless access points to local customers and shared the Internet connection for a monthly fee. The Mali station set up a cybercafé at the radio station (Farm Radio International 2011a).

Lessons from the AFRRI Project

Production and Outreach

Many of the skills and techniques involved in PRCs are likely to be new to broadcasters. FRI has developed week-long courses accompanied by training materials to help African radio broadcasters learn techniques they can use in PRCs. FRI is now developing an e-course to provide broadcasters with tools to design effective and engaging agricultural programs (support for the e-course is from the Commonwealth of Learning: see http://ecourse.farmradio.org/).

FRI is also producing farm radio resource packs (FRRPs) available online that will focus on African agricultural value chains defined as "the people and activities that bring a basic agricultural product … from production in the field to the consumer, through stages such as processing, packaging, and distribution." The resource packs will include information documents, sample radio scripts, materials on key issues, and other resource materials (see http://www.farmradio.org/radio-resource-packs/). These training materials and courses could be adapted to meet the needs of broadcasters in other developing regions who could put these concepts to use in other settings.

Training in Use of ICTs

As AFRRI points out: "Technology on its own cannot improve farm radio's reach. Introduction of ICTs must be integrated with appropriate training on the use and maintenance of equipment, combined with sustainable ways for broadcasters to fund and own the technology" (Farm Radio International 2011a). Most broadcasters will need training in technologies and software which are new to them such as digital editing software, point-to-multipoint SMS, and IVR systems that were introduced in the AFRRI project. However, broadcasters will also likely need training in how to use equipment that they are familiar with such as mobile phones, MP3 recorders, and Internet access to apply them for participatory media campaigns.

Capital Costs and Operating Costs

In the AFRRI project, participating radio stations were given the equipment they would be using during the project period and told that they would be able to keep the equipment afterwards. In some cases, this approach of donating equipment can lead to lack of concern about taking care of equipment or replacing worn-out or damaged parts. However, AFRRI researchers reported that ownership contributed to a sense of responsibility for the equipment and encouraged some to explore innovative ways of using the technology. AFRRI concluded: "When the station believes that the equipment belongs to them and not to an NGO or the government, and they see its value, they are more likely to take it upon themselves to seek solutions and fix problems" (Farm Radio International 2011a).

Of course, most radio stations would have to find the funds to buy their own equipment. But in the long run, the operating costs are likely to be much more of a significant challenge, whether or not the participants had up-front funding for equipment. For example, the author found in evaluating telecenter projects that some managers did not budget to replace printer cartridges, although printing, photocopying, and desktop publishing were often the main revenue generators for the telecenter (Hudson 2001). In the AFRRI project, one of the ongoing costs was for connectivity—for Internet access and for mobile phone airtime for callouts, call-ins, SMS reminders, and staff coordination. Although mobile phones turned out to be very cost effective, airtime in many African countries can be expensive relative to salaries and other operating costs.

Appropriate Technology

Identifying criteria for equipment selection that will help to minimize costs but also assure usability can be important in minimizing overall equipment costs while maximizing benefits. These criteria will vary depending on many local factors including equipment availability and pricing, physical conditions where equipment will be used in the studio and in the field, and ICT literacy of staff and of other users.

In the AFRRI project, considerations included availability of equipment and parts, serviceability, affordability, and simplicity of use. There are numerous examples in the developing world of donated equipment lying unused because spare parts are not available or prohibitively expensive to import, or no one has the technical expertise to fix the equipment. The AFRRI project attempted to procure equipment that was both durable and repairable locally; AFRRI also provided technical training to those who would be using the equipment.

Sustainability and Entrepreneurship

Small and nonprofit radio stations face the challenge of ongoing sustainability to cover their operating costs, regardless of whether they receive donated equipment or supplies. Of course, paid advertising and announcements are well-known means of generating revenue, but they may not be appropriate for nonprofit stations or generate much revenue in small markets. The AFRRI project explored some innovative approaches to generate revenue using ICTs:

- *On-air classified ads*: A radio program in Tanzania announces classified ads received throughout the day by SMS messages sent to the broadcaster's phone. The customer who wishes to place an ad on the radio program must pay 2,000 Tanzanian shillings (about $US1.25). The payment is actually made by sending mobile phone airtime credit to the broadcaster's phone. Once the ad and the payment are received, the ad is read on the air.
- *On-air greetings via SMS*: Another Tanzania radio station charges for messages such as prayers and greetings sent to family and friends around the country.

A person who wants a message read on the air pays for a premium SMS (approximately $US.30) sent to a special number at the radio station. The revenue from this SMS is then divided between the radio station and the mobile phone company (Farm Radio International 2011a).

- *Hosting IVR services*: A Ghanaian radio station that continued to provide IVR services which farmers could access to get information on demand found that other organizations and businesses became interested in offering IVR services. Freedom Fone technology used for AFRRI's IVR experiments allows radio stations to host and support IVR services virtually by assigning a unique SIM card for each local client. The station planned to develop other value-added services such as co-designing the IVR for clients, supplying information for the audio menus, recording/producing the audio menus, and offering discounts on radio spots to advertise clients' IVR service and phone number (Farm Radio International 2011a).
- *Cybercafé at the radio station*: A radio station set up an Internet café on its premises and used the revenue to cover its own Internet costs and pay an in-house technician.
- *Wireless Internet service provider (WISP)*: A radio station may be able to become a WISP by setting up a Wi-Fi transmitter and selling access to its Internet connection. For example, an AFRRI radio station used wireless access points to make its Internet connection available to surrounding businesses, government offices, and schools. This cost-sharing model enabled the stations to cover the $US250 monthly Internet fees plus additional revenue to pay technical staff at the radio station. (However, reselling Internet access may not be legal in some jurisdictions.)

The Potential of Video Broadband

In industrialized countries, television broadcasting was the initial means of disseminating video content to households. Mobile phones brought voice and text messages, and smartphones are becoming platforms for video. However, in rural Africa access to broadcast television remains limited because over-the-air signal coverage outside of cities is often poor and a majority of rural Africans still lack electricity. TV sets must be run from generators or solar panels that require conversion to AC current. But mobile phone penetration has increased dramatically so that even in low-income rural areas, access is almost universal (although relatives may share phones). As smartphones become cheaper, they are likely to become the most accessible platform for video in rural Africa.

In addition to providing news and entertainment, mobile video could be a powerful tool for rural development. Much as audio MP3 players were used with AFRRI, community members could record video clips for others to view and could see examples of farming practices and stories of successful farmers and entrepreneurs. The AFRRI project also introduced IVR so that farmers could use their mobile phones to seek information using audio menus. However, video could be much more effective in illustrating responses.

Also YouTube has become an important repository for training materials. Indigenous broadcasters in northern Canada who make training materials in native languages now post the videos online on YouTube so that they are available on demand for all communities in the region (Hudson 2011). This approach could be very useful in Africa for agricultural development as well as for public health education and vocational instruction. Many of the lessons of the AFRRI project are relevant for such video applications.

Mobile broadband could also extend the reach of Internet content beyond cyber cafés that are available only in towns and schools where access is often limited to students, and facilities are typically available only during school hours.

However, the price of connectivity is likely to be the most significant constraint. Rural residents use phone cards that require prepayment for voice calls and text messaging. The price of these basic services often forces rural residents with very limited cash incomes to go without mobile service until they can afford to recharge their phone cards. Even where mobile networks are being upgraded to carry broadband, their use will be severely constrained if current pricing models from the industrialized world apply.

Mobile Broadband for Development

The AFRRI project has shown that radio plus interactive ICTs can increase not only awareness but also adoption of farming practices to improve food security in rural Africa. Access to video, most likely through mobile broadband, can build on AFRRI experience with the potential to provide relevant information to farmers with limited formal education, and while enabling them to exchange their own video content to enhance social entrepreneurship.

References

Farm Radio International. (2011a). *The new age of radio: How ICTs are changing rural radio in Africa*. Ottawa, ON: FRI.

Farm Radio International. (2011b). *Participatory radio campaigns and food security: How radio can help farmers make informed decisions*. Ottawa, ON: FRI.

Food and Agriculture Organization. (2012). *The state of food insecurity in the World 2012*. Rome: FAO.

Girard, B., et al. (2003). *Radio, new ICTs and interactivity*. Rome: FAO.

Hudson, H. E. (2001). *Telecentres for development: Issues and strategies. Telecentres for Development*. Vancouver, B.C.: Commonwealth of Learning.

Hudson, H. E. (2006). *From rural village to global village: Telecommunications for development in the information age*. New York: Routledge.

Hudson, H. E. (2011). Rural broadband: Strategies and lessons from North America. *Intermedia, 39*(2).
McAnany, E. (2012). *Saving the world: A brief history of communication for development and social change*. Urbana, IL: University of Illinois Press.
Ndiaye, M., et al. (2008). *Radio and ICT in West Africa: Connectivity and use*. Dakar: Panos Institute West Africa.
Schramm, W. (1964). *Mass media and national development: The role of information in developing countries*. Palo Alto, CA: Stanford University Press and UNESCO.
Ward, D. (2010). *Manual for participatory radio campaigns*. Ottawa, ON: Farm Radio International.

Chapter 11
The Economic Impact of Telecommunications in Senegal

Raul L. Katz and Pantelis Koutroumpis

Introduction

Ever since the launch of the Internet, but especially in the last 10 years, researchers have been attempting to measure the economic contribution of communications. Originally, the approach focused on the study of cross-sectional samples of countries. Due to limitations on data availability, the primary emphasis had been on OECD countries (facilitated by the extensive Eurostat data sets) or worldwide analysis (based on ITU statistical indicators). While this approach has continued to be pursued,[1] researchers have focused recently on quantitative assessment of specific country studies. For example, aiming to understand the economic impact of broadband, the authors have conducted studies in Europe (Katz et al. 2010a), the United States (Katz and Suter 2009; Katz et al. 2010a, b), and Latin America (Katz 2011c; Katz 2010; Katz et al. 2011; Katz and Callorda 2011).

Increased data availability and lengthier time series would now allow researchers to extend single-country analyses by developing tracking models. Adding to cross-country comparisons and single-country studies, research can now shed light on how is technology contributing to the economy within a single

The following study was funded by the General Secretariat of Orange with the support of Orange Senegal. The views expressed in this chapter are those of the authors and do not necessarily reflect the opinions of Orange.

[1] See Koutroumpis (2009), Waverman (2009), and Katz (2012).

R.L. Katz (✉)
Director, Business Strategy Research, Columbia Institute for Tele-Information (CITI), New York, NY 10027, USA

Adjunct Professor, Finance and Economics Division, Columbia Business School, New York, NY 10027, USA
e-mail: rk2337@columbia.edu

P. Koutroumpis
Fellow, Imperial College, Business School, South Kensington Campus, London SW7 2AZ, UK

J. O'Neill et al. (eds.), *Broadband as a Video Platform*, The Economics of Information, Communication, and Entertainment, DOI 10.1007/978-3-319-03617-5_11,

country over time. Policy makers could greatly benefit from this expanded view since it would allow them to change course or fine-tune policies without relying only on a "rearview" mirror or an analysis of "expected" impact.

The following study represents a contribution to this new capability. In 2010, the authors conducted an assessment of the economic impact of telecommunications in Senegal, focusing on the relative impact of wireless and broadband communications. This new study, which enhances time series with data up to 2013, attempts to answer three questions raised as hypotheses 2 years ago.

First, the 2010 study concluded that telecommunications had a sizable *direct* economic impact, accounting for over 10 % of Senegal's gross domestic product (GDP). In fact, Senegal's National Accounts indicated that between 2002 and 2010, the rate of growth of telecommunications' contribution to Senegal's GDP was three times that of construction and financial services, generating as much (Francs CFA 419 billion) as the two other sectors combined. Since 2010, Senegal's economy growth has slowed down to 2.6 % in 2011 (from 4.2 % the year before) but increased in 2012, reaching 3.7 %. While inflation reached 3.38 % driven by high energy prices, it dropped to 1.42 % in 2012. Finally, the rate of Gross Capital Formation increased from 28.93 % in 2010 to 30.33 % in 2011. In this context of changing economic conditions, it is relevant to research how sustainable has telecommunications' direct economic contribution been to the Senegalese economy. Is volatility in the economy as a whole mirrored by changes in telecommunications' contribution to the GDP, or is the sector continuing to grow anticipating a structural shift in composition of output?

Second, the econometric modeling of the 2010 study generated evidence of a substantial *indirect* contribution of wireless telephony. According to that study, the annual contribution on GDP growth from mobile phones was approximately 0.044 % for every 1 % of increase in wireless penetration. Given that the economy had grown—on average—at 4.1 % between 2003 and 2010, this figure suggested that wireless alone was responsible for 13.6 % of all economic growth in Senegal for that period. The study also concluded that, given the future growth in the penetration of wireless, the impact of the technology on GDP would continue to increase, albeit at a slower rate due to a saturation effect. As expected, wireless telephony has continued to grow in Senegal at a fast pace, averaging 18 % annually. Wireless penetration (in connections) has increased from 66.22 % by the end of 2010 to 92.68 % two and a half years later. Could we, in light of market developments, test the hypothesis of "declining returns" to wireless penetration? Or, alternatively, are we witnessing a "return to scale" whereby impact increases with penetration? This last counter-hypothesis is not to be discarded since universal adoption of the technology could lead to the creation of new markets and the acceleration of innovation.

Third, the study also did not find any econometric evidence of an economic contribution from broadband in Senegal for the period 2004–2011. This result was not inconsistent with the findings of other studies, where in many emerging countries broadband penetration yields almost nil economic impact due to very low network coverage, a concentrated market structure, consequent higher prices, and limited consumer interest due to minimal local content and applications. However, the study also anticipated that the future of broadband in Senegal, particularly with regard to mobile broadband, could change this situation. Wireless broadband has, indeed, undergone a substantial growth in the past 2 years. While by the end of 2010 mobile broadband represented 24,000 subscriptions and fixed broadband amounted to 78,000, 2 years later mobile broadband subscriptions reached 491,000, compared to 95,000. Mobile broadband has continued to grow exponentially, reaching 631,000 by mid-2013. Could we actually start to witness the emergence of broadband's economic impact as a result of the dramatic growth in mobile broadband?

In this context, a study that revisits the impact equation 2 years later is quite pertinent for several reasons. First, how sustainable has the direct economic contribution of telecommunications been? Furthermore, even if the direct contribution did not change substantially, would it be possible that the indirect impact increases? Second, is there a condition of "declining returns" to wireless telephony, whereby it would be imprudent to extrapolate economic effects from mid-penetration levels to saturation points? Or, alternatively, is there a growing "return to scale" where economic effects tend to increase exponentially with penetration? Third, can we find evidence of economic effects of mobile broadband that would lead governments in emerging countries to formulate policies that stimulate its growth to the detriment, perhaps, of fixed broadband?

Along these lines, a caveat should be made. The measureable economic impact of ICT infrastructure depends heavily on the introduction timing, existing adoption conditions, and market maturity. As studies of the lagged impact of ICT have demonstrated (Hardy 1980; Jorgenson et al. 2007; Karner and Onyeji 2007), significant economic effects of ICT do not materialize immediately after the introduction of a new technology. This study is predicated on the assumption that 2 years (or maybe less) could be sufficient enough to start detecting economic effects. An underlying premise of the research is that given the speed with which technology is developing around the world, multi-year lags in assessing social and economic impact might not be necessary, and that, measurement techniques notwithstanding, policy makers are better served by conducting multiple and frequent tracking studies that allow them to monitor the emergence of new effects and accelerate, slow down, or correct policies.

This study begins by providing a brief review of the results of last year's study ("Summary of 2010 Study Results" section). It then presents the changes that have taken place both in the Senegalese economy and in the telecommunications industry since last year ("Changes in the Senegalese Economy and the Telecommunications Sector Since 2010" section). This provides a context for conducting a new iteration of econometric modeling with the additional observations ("New Study Results" section). The implications from a public policy standpoint are drawn in "Discussion of Study Results" section.

Summary of 2010 Study Results

The impact of telecommunications in the period ending in 2010 on the Senegalese economy was assessed first in terms of the sector's direct impact, resulting from its importance in the GDP, the employment being generated by its operators and their local suppliers, and taxes being paid. Secondly, the economic impact of telecommunications was also assessed in terms of its indirect "spill-over" impact on the economy as a whole, by contributing to the growth of the GDP across sectors.

Direct Economic Contribution of Telecommunications Until 2010

The total revenues of the Senegalese telecommunications industry represented 1.4 billion USD (10.8 % of the national GDP) in 2010. This ratio had been consistently growing since 2002 indicating the increasing importance of the sector (see Fig. 11.1).

Furthermore, the value-added of the telecommunications and postal services sectors had reached 850 million USD, while their contribution to GDP growth was 104 million USD (8.5 %), the highest across industrial sectors.[2,3] When comparing it against other sectors, the direct contribution to GDP growth of the telecommunications and postal services sectors was higher than energy, construction, and finance (approximately 26 million USD each sector) and had been consistently growing faster since 2002 (see Fig. 11.2).

In parallel to its direct economic contribution, the telecommunications industry had an important impact in the creation of direct jobs (i.e., telecommunications employment). The total number of direct jobs in telecommunications reached 3,000, while the indirect employment reached 55,000. The total direct and indirect employment figure represented in 2010 1.11 % of the total employment in Senegal and 10.6 % of the service sector.[4] Finally, the Senegalese telecommunications sector contributed in 2010 12.6 % of the public treasury in terms of taxes being paid. For example, Orange Sonatel, the incumbent operator, contributes 10.4 % of the total fiscal revenues of the country.[5]

---

[2] When considering it with transportation (*source*: National Accounts).

[3] GDP of Senegal in 2010 was $12,954 m (current).

[4] Total population employment reached 5,194,107 in 2010; according to the Youth Employment Network study on Ghana and Senegal, 10.5 % of the labor force is employed in the service sector.

[5] *Source*: Sonatel.

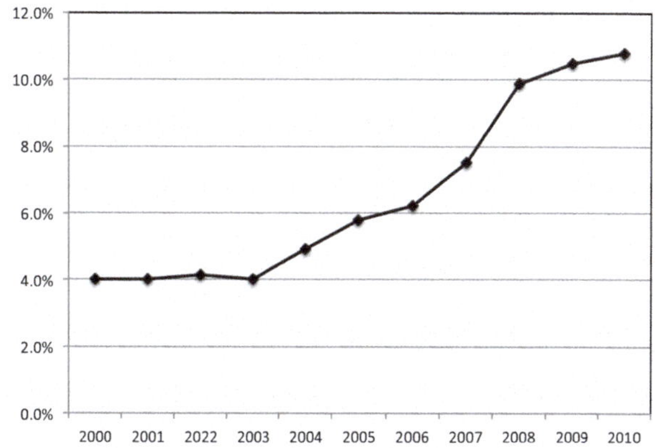

Fig. 11.1 Percentage of GDP. *Sources*: IMF; World Bank

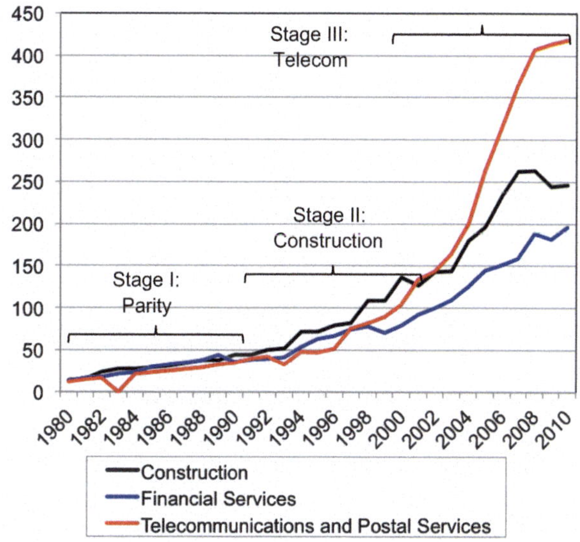

Fig. 11.2 Senegal: GDP by industry (1980–2010) (in CFA billion). *Source*: Senegal National Accounts

Indirect Economic Contribution of Telecommunications Until 2010

As indicated above, beyond the direct economic contribution, telecommunications can have a positive indirect contribution to economic growth. Given the different penetration rates exhibited by mobile telephony and broadband, the analysis of economic impact of both technologies was conducted separately.

Based on structural econometric models, mobile telephony was found to significantly affect the Senegalese economy between 2003 and 2010. The annualized average contribution to the GDP was estimated to be equal to 0.044 % growth of GDP for every 1 % increase of mobile penetration (see Table 11.1).

Table 11.1 Results of 2010 mobile telephony model

Variables	Mobile model
Growth (GDP_{it})	
Labor force (L_{it})	0.416***
Fixed capital stock (K_{it})	0.615***
Mob penetration (Mob_Pen$_{it}$)	0.044*
Constant	–
Demand (Mob_Pen$_{it}$)	
GDPC ($GDPC_{it}$)	0.165
Mob. price ($MobPr_{it}$)	−5.238***
Market conc (HHI_{it})	−3.590***
Constant	10.588***
Supply (Mob_Rev$_{it}$)	
Mob price ($MobPr_{it}$)	−3.122***
GDPC ($GDPC_{it}$)	0.929***
Market conc (HHI_{it})	0.123
Constant	−3.360***
Output (ΔMob_Pen$_{it}$)	
Mob revenue (Mob_Rev$_{it}$)	0.867***
Constant	7.150***
Year effects	YES
Quarter effects	YES
R^2	*(1)*
Growth	0.99
Demand	0.98
Supply	0.98
Output	0.30

* Significant at 80 %
** Significant at 90 %
*** Significant at 99 %

These results were found to fit rather well in the logarithmic growth impact curve of Gruber and Koutroumpis (2011). With a median mobile penetration of approximately 35 % between 2004 and 2010, every 1 % increase in mobile penetration would yield a 0.044 % increase in GDP. This estimate was only 3 % lower than the estimate of Gruber and Koutroumpis' exponential model[6] (see Fig. 11.3).

Based on the prior model, the expected impact of the sector according to industry forecasts was also estimated. Since it was estimated that mobile penetration would reach 115 % by the end of 2016 suggesting a saturation stage, the median mobile penetration for the period of study (2004–2016) would increase to 61.4 % and the annual impact from each 1 % increment of mobile penetration would contribute slightly more than 0.05 % to GDP. In fact, we hypothesized at the time that the wireless market saturation would lead to a stagnation of the economic effect of mobile adoption over time and that while the growth effects would be preserved, they would be transferred to broadband.

[6] Estimated value: $(0.0074 \times \ln(0.35)) + 0.0533 = 0.0455$, actual estimate: 0.44.

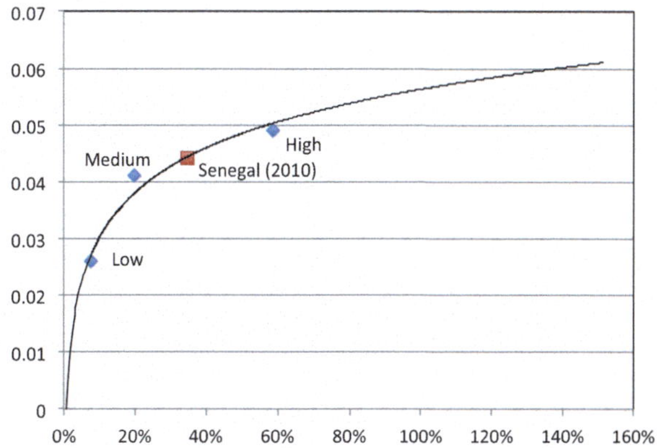

Fig. 11.3 Estimate based on fitted line of previous studies (median mobile penetration ~35 %)

For the analysis of the impact of fixed[7] broadband on the economy a model similar to the mobile telephony structural model was utilized. According to the model, no significant effects were found from the adoption of broadband for the period 2004–2010 (see Table 11.2).

This result was not inconsistent with the findings yielded by other studies such as the one completed by one of the authors for Colombia (Katz et al. 2011), where broadband penetration of 4.83 % yielded an economic contribution of 0.003 % to GDP growth for every 1 % increase in penetration. We hypothesized, however, that the real growth potential of broadband would come from the adoption of mobile broadband in the country.

Changes in the Senegalese Economy and the Telecommunications Sector Since 2010

Changes in the Senegalese Economy Since 2010

The 2010 study was conducted at the time where the global recession in 2009 and the Senegalese food and energy crisis of 2008 had significantly affected the country's economy. In fact, recovery began in 2010 after a combination of domestic structural reforms and the improved results of the global economy. GDP growth increased from 2.2 % in 2009 to 4.2 % in 2010, although it fell back to 2.6 %[8] in 2011. This number recovered in 2012, reaching 3.7 %, with projected growth of 4.3 and 5.1 % for 2013 and 2014, respectively. These projections are based on future

[7] Mobile broadband adoption was embryonic at the time.

[8] IMF: http://www.imf.org/external/pubs/ft/weo/2012/01/index.htm.

Table 11.2 Results of 2010 broadband impact model

Variables	Broadband model
Growth (GDP_{it})	
Labor force (L_{it})	0.402***
Fixed capital stock (K_{it})	0.552***
Broadband penetration (BB_Pen_{it})	−0.003
Constant	–
Demand (BB_Pen_{it})	
GDPC ($GDPC_{it}$)	0.832**
BB. price ($BBPr_{it}$)	−0.794***
Education (Edu_{it})	0.082
Urbanization (URB_{it})	25.402***
Constant	−87.929***
Supply (BB_Rev_{it})	
BB price ($BBPr_{it}$)	0.161
GDPC ($GDPC_{it}$)	3.273***
Constant	−7.223***
Output (ΔBB_Pen_{it})	
BB revenue (BB_Rev_{it})	0.572
Constant	7.554
Year effects	YES
Quarter effects	YES
R^2	*(1)*
Growth	0.99
Demand	0.99
Supply	0.35
Output	0.16

implementation of the government's socioeconomic program and compliance with the IMF's Policy Support Instrument.[9]

The start of the recovery is also depicted by the changes in inflation. After a sharp rise during 2005–2008, inflation turned negative reaching −0.98 % in 2009 and 1.37 % in 2010. In 2011 inflation reached 3.38 % mainly as a result of higher energy prices but dropped down to 1.42 % in 2012[10] and hovered at 1.1 % until July 2013[11] (see Fig. 11.4).

Total imports of goods and services had steadily grown after 2001 reaching a peak of 53.8 % of GDP in 2008 but experienced a sudden drop thereafter declining to 44 % of GDP in 2010. Exports further fell to 24 % in 2011 further expanding the external balance of goods and services (see Fig. 11.5).

Foreign Direct Investment dropped from 3.01 to 1.62 % of GDP between 2008 and 2009, reaching 1.98 % in 2011. The rate of investment measured by the gross fixed capital formation (GFCF) was positive: 27.97 % of GDP in 2009, 28.93 % in

[9] African economic outlook: http://www.africaneconomicoutlook.org/en/countries/west-africa/senegal/.

[10] World Bank.

[11] http://www.tradingeconomics.com/calendar.

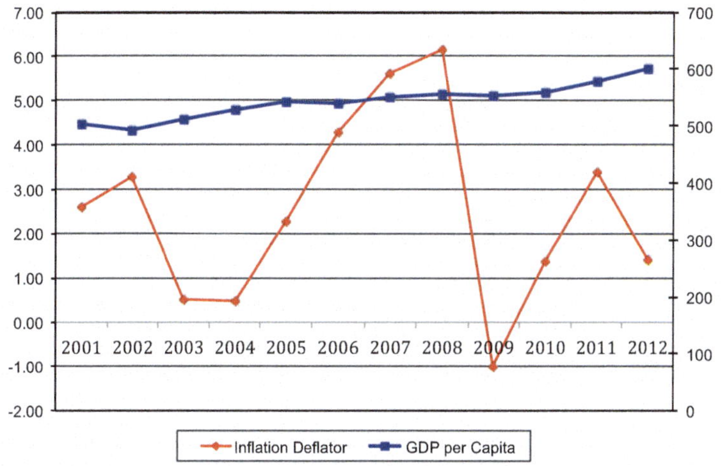

Fig. 11.4 GDP per capita and inflation. *Source*: World Bank (2011)

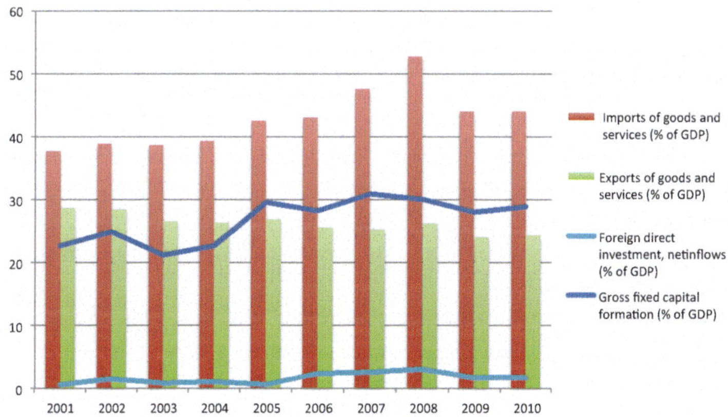

Fig. 11.5 Imports, exports, gross fixed capital formation, and foreign direct investment as % of GDP. *Source*: World Bank (2011)

2010, and 30.3 % in 2011. It is forecast that Senegal's external position will gradually improve during 2013 (7.1 %) and 2014 (6.7 %) after the steep increase to 7.6 % in 2011.[12] The overall budget deficit, including grants, increased from 5.2 % of GDP in 2010 to 6.7 % in 2011 but decreased to 5.9 % in 2012. It is estimated to drop to 4.9 % in 2013 and 4.3 % in 2014.

In light of these macroeconomic trends, it would be relevant to explore whether the relationship between the sector and the overall economy has changed at all.

[12] The 2011 IMF debt sustainability analysis (DSA).

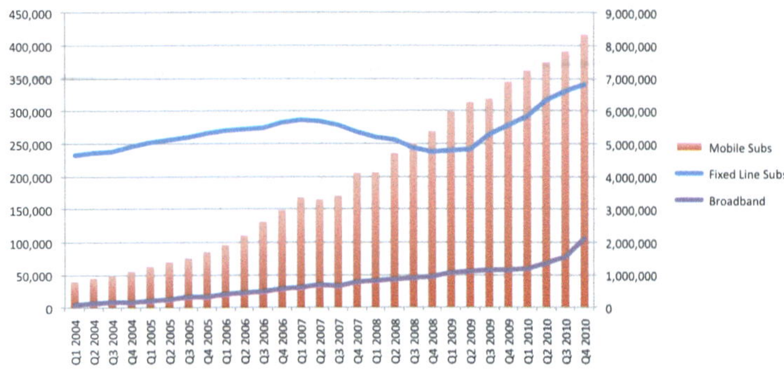

Fig. 11.6 Mobile, fixed line, broadband, and 3G subscribers in Senegal. *Source*: ARTP Senegal (2011)

Has the elasticity between GDP growth and telecommunications output changed? Has the sector benefitted from the increase in Gross Capital Formation?

Changes in the Telecommunications Industry Since 2010

As reported in the 2010 study, the telecommunications sector had grown rapidly during the last 15 years in Senegal primarily driven by the adoption of mobile telephony (see Fig. 11.6).

Since 2010, wireless has continued its growth, albeit with a momentary slowdown in 2011 (see Table 11.3).

In addition to the dramatic increase in wireless subscribership, fixed lines have also increased since 2008. Fixed lines had experienced a very slow rate of adoption before 2007 reaching 286,000 lines. Significant substitution effects took place after the quick adoption of mobile services resulting in 21 % drop in fixed line services during 2007–2009. This 3-year drop in fixed line subscribers lasted until the second quarter of 2009. A steep rise in the demand of ADSL lines after 2009 has turned this trend, with fixed lines reaching an all-time high in 2011 (346,400 lines), before declining slightly in 2012 to 338,167. This phenomenon of simultaneous growth of both technologies indicates an industry context of a market searching for any possibility to meet its needs through either technology. While the growth rate in mobile telephony is dramatically higher than wireline, the Senegalese market continues to exhibit less of a technology substitution dynamic as the one that can be seen in other emerging countries.

The most important change since 2010 in the telecommunications market has been the development of mobile broadband. While ADSL lines continue to grow, the most significant boost in the Senegalese broadband market has been the deployment and launch of 3G networks that now account for 80 % of broadband connections (see Table 11.4).

Table 11.3 Wireless connections and subscribers in Senegal. *Source*: ARTP Senegal; GSMA Intelligence; IMF; analysis by the authors

	2008		2009		2010		2011		2012	
	Value	Annual delta	Value	Annual delta	Value	Annual delta	Value	Annual delta	Value	Annual delta
Wireless connections	5,389,133	48.43 %	6,901,492	28.06 %	8,343,717	20.90 %	9,325,828	11.77 %	11,470,996	23.00 %
Market penetration	45.72 %		57.00 %		67.10 %		72.93 %		87.51 %	
Wireless subscribers	3,467,227	34.57 %	4,165,843	20.15 %	4,778,799	14.71 %	5,208,366	8.99 %	5,955,032	14.34 %
Market penetration	29.02 %		33.95 %		37.93 %		40.26 %		44.84 %	

Note: The difference between connections and subscribers is driven by double SIMs

Table 11.4 Subscriber market shares across telecommunication platforms in Senegal (2010). *Source*: IMF; Orange; ITU; analysis by the authors

		2009		2010		2011		2012	
		Number	Annual growth rate	Number	Annual growth rate	Number	Annual growth rate	Number	Annual growth rate
Fixed broadband	Number	58,720	23.99 %	78,647	33.94 %	92,713	11.88 %	95,561	3.07 %
	Penetration	0.49 %		0.63 %		0.73 %		0.73 %	
Wireless broadband	Number	5,067		35,591	602.41 %	188,362	429.24 %	447,786	137.73 %
	Penetration	0.04 %		0.29 %		1.48 %		3.42 %	

Table 11.5 Subscriber market shares across telecommunication platforms in Senegal (2010). *Source*: Orange Sonatel; Business Monitor International; analysis by the authors

	Fixed line		Fixed broadband		Wireless	
Player	2010	2012	2010	2012	2010	2012
Orange Senegal	92.4 %	86.4 %	97.41 %	95.00 %	60.41 %	62.85 %
Tigo Senegal					27.96 %	24.30 %
Expresso	7.53 %[a]	13.6 %	2.59 %	5.00 %	11.63 %	12.85 %
Total	100 %	100 %	100 %	100 %	100 %	100 %

[a]Fixed wireless offering

From an industry structure standpoint, the mobile market was still a monopoly of the local incumbent—Sonatel, now Orange Senegal—until 1999 when a second operator—Tigo—entered. In 2009, Expresso, the third operator, started to provide services in the Senegalese market. While the market is still primarily controlled by Sonatel (62.85 % in 2012), Tigo has 24.30 % of the customer base and Expresso the remaining 12.85 % (see Table 11.5).

Despite the apparent stability of market shares, competitive intensity in the wireless market has increased since 2010. The Hirschman–Herfindahl Index (HHI), that measures market concentration, declined from 4,634 in 2010 to 4,603 at the end of 2012, reaching 4,506 by mid-2013. An assessment of market share in wireless by specific products highlights the centers of gravity of the competitive battles (see Table 11.6).

As Table 11.6 indicates, competition as noted by market share volatility has been intense across all segments of the wireless industry. Since 2010, Tigo has lost five points of prepaid share at the expense of Orange and Expresso, Orange has lost five points of postpaid share to the other two players, and the wireless broadband market is now being fought between Orange and Expresso.

Finally, total service revenues have increased US$ 100 million since 2010, reaching a total of US$ 1.5 billion. The trend clearly relates to the rise in mobile adoption and usage as well. The slight increase since 2010 does not match the previous momentum between 2002 and 2008 (see Fig. 11.7).

As a result, the elasticity of telecommunications demand with respect to the growth of the economy is still behind that experienced between 2001 and 2008 (see Fig. 11.8).

This has to be considered in the context of the dramatic growth of mobile telephony. In fact, the stabilization of revenue growth is primarily due to the decline in mobile average revenue per user (ARPU) resulting from increased competition. Nevertheless, the data shows that the telecommunications sector growth is now aligned with the GDP growth trend. We will come back in "Discussion of Study Results" section to provide an interpretation of this trend.

Table 11.6 Subscriber market shares across telecommunication platforms in Senegal (2010). *Source*: GSMA Intelligence

	Prepaid		Postpaid		Wireless broadband	
Player	2010	2012	2010	2012	2010	2012
Orange Senegal	60.89 %	61.99 %	77.44 %	72.45 %	100 %	68 %
Tigo Senegal	28.35 %	23.08 %	11.55 %	13.64 %		
Expresso	10.75 %	14.93 %	11.02 %	13.91 %		32 %
Total	100 %	100 %	100 %	100 %	100 %	100 %

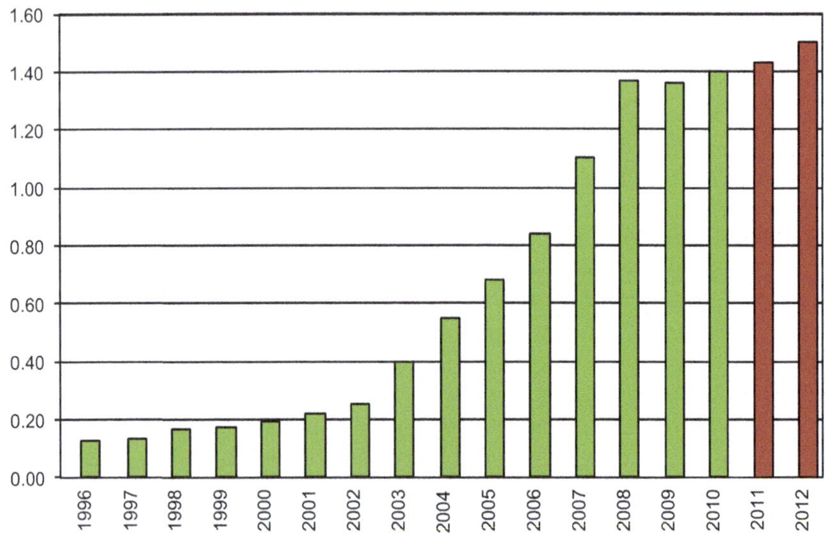

Fig. 11.7 Telecommunications service revenues 1996–2010. *Sources*: ITU; Euromonitor; The Economist; analysis by the authors

New Study Results

To what extent do the changes both in the economy and the telecommunications sector changed the economic impact equation? This will be explored in terms of descriptive economic statistics similar to the ones reported in the first study for the direct effects as well as equivalent econometric modeling techniques for the indirect contribution.

Direct Economic Contribution Since 2010

Two years after the 2010 study, the total revenues of the Senegalese telecommunications industry increased by US$ 100 million (slightly dropping as a percent of

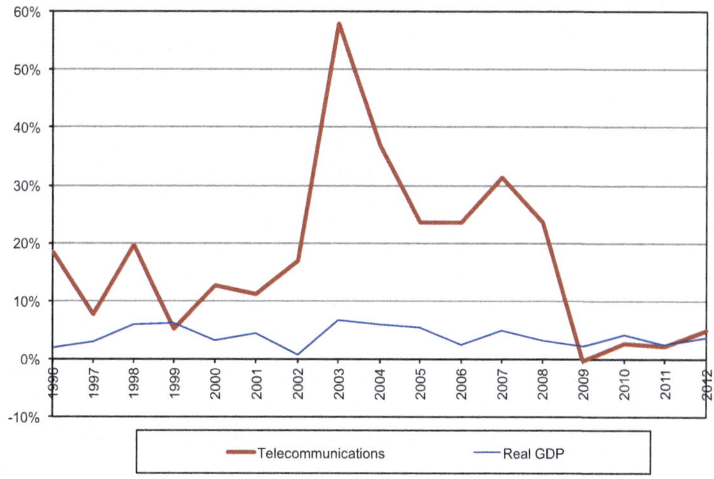

Fig. 11.8 Annual change in real GDP and the telecom market (1996–2012). *Sources*: ITU; World Bank; IMF; ISI; analysis by the authors

the GDP from 10.8 % in 2010 to 10.3 % in 2011 and increasing to 10.6 % in 2012).[13] This ratio, which had been consistently growing since 2003 indicating the increasing importance of the sector , has been relatively stable since 2009 (see Fig. 11.9).

Furthermore, the output of the telecommunications and postal services sectors reached Franc CFA 438 billion, still higher than construction and finance, but losing ground to the two other sectors given their rebound in 2011 (see Fig. 11.10).

Indirect Economic Impact Since 2010

To measure the indirect economic impact since 2010 we constructed a model similar to the one specified for the 2010 study for mobile telephony. To reiterate, the model consists of four equations: an aggregate production function modeling the economy and, subsequently, three demand, supply, and output functions. The last three functions model the wireless market operation, and, controlling for the reverse effects, the actual impact of the infrastructure is estimated. In the production function, GDP is linked to the fixed stock of capital, labor, and the mobile infrastructure proxied by mobile penetration. The demand function links mobile penetration to the average consumption propensity of individuals proxied by GDP per capita, the cost of a basic mobile service, and the competition in the mobile market, measured by the HHI index. The supply function links the aggregate mobile

[13] National Accounts are still not published for 2012. Data for 2012 sector revenues is estimated by Hot Telecom Services.

- Capital deepening has an unusually high impact on economic growth (coefficient of 0.785 versus 0.1349 for Cote d'Ivoire).
- Incomes are crucial for adoption and investments (coefficient: 5.635), which indicates that affordability remains a critical barrier for demand.
- Competition has a positive impact on investments (coefficient: −0.0005 and significant), but not on adoption (not significant).

The actual contribution of mobile technology was calculated by multiplying the compound annual growth rate of wireless penetration between 2005 and 2013 (11.5) by the coefficient of economic impact derived from the econometric model presented in Table 11.7 (11.6):

$$\begin{aligned} \text{CAGR} = {} & (\text{Wireless penetration 2Q 2013}\,(92.68\%) \\ & -\text{Wireless penetration 4Q 2005}\ \ (15.70\%))^{(1/7.5)-1} \end{aligned} \tag{11.5}$$

The CAGR for Senegal wireless telephony for the period 2005–2013 is 26.71%.

$$\begin{aligned} \text{Impact of wireless on GDP (2005–2013)} = {} & \text{CAGR}\ (26.71\%) \\ & \times \text{Coefficient of Impact}\ (0.061) \end{aligned} \tag{11.6}$$

According to the formula, the annual contribution to GDP from mobile phones is 1.63 % of GDP. Based on the difference between the 2012 GDP of US$ 14,159 million and the 2005 GDP of US$ 8,699 million, the indirect annual contribution of wireless telephony amounts to US$ 176 million.

Furthermore, looking at the results in light of the exponential growth impact curve of Gruber and Koutroumpis (2011), the impact forecast in 2010 study is confirmed. To reiterate, in 2010, with a median mobile penetration of approximately 35 % in the sample period, the country had a coefficient of 0.044. Next, considering that mobile penetration was forecast to reach 115 % by the end of 2016, the median mobile penetration for the period of study (2004–2016) would have shifted to 61.4 %, resulting in a coefficient of 0.05 %. Now, with median mobile penetration in the 2004–2012 period shifting to 61.4 %, the coefficient has increased to 0.061 indicating an acceleration of impact (see Fig. 11.11).

The new coefficient being higher than that estimated by Gruber and Koutroumpis' exponential model could be partially explained by the standard deviation of the specification. However, another potential explanation for which we remain cautious about is that the economic contribution of mobile telephony is accelerating due to the new services that rely on telecommunications to reach consumers. We consider, in particular, financial services like mobile money that rely on text messaging.

For the analysis of the impact of mobile broadband on the Senegalese economy a model similar to the mobile telephony structural model was utilized. The model also consists of four equations: an aggregate production function modeling the operation of the economy and subsequently three demand, supply, and output functions. The latter functions model the mobile broadband market operation and

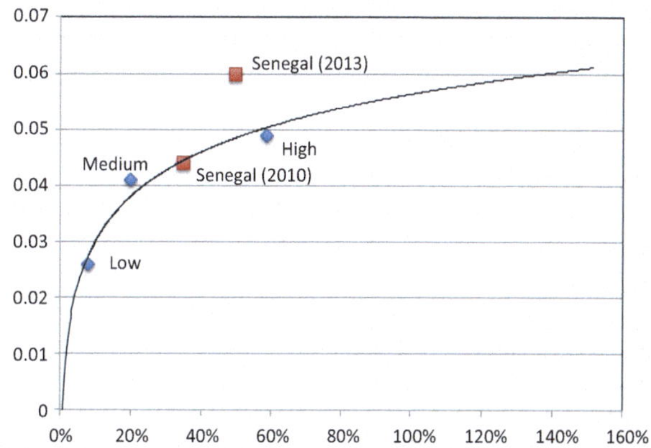

Fig. 11.11 Estimate based on fitted line of previous studies (median mobile penetration: 2004–2010: ~35 %; 2010–2016: 61.4 %)

estimate the economic impact of mobile broadband while controlling for the reverse effects. The demand function links mobile broadband penetration to the average consumption propensity of individuals proxied by GDP per capita, the cost of a basic mobile broadband service (price of a monthly subscription), the percent of individuals that fulfill secondary education, and the percent of population residing in densely populated urban areas. The supply function links the aggregate mobile broadband revenue to the relevant price levels and the GDP per capita. The infrastructure equation links annual change in mobile broadband penetration to the market revenues, used as a proxy of the capital invested in a country during 1 year.

The econometric specification of the model is as follows:

Aggregate production function:

$$\mathrm{GDP}_{\mathrm{it}} = a_1 K_{\mathrm{it}} + a_2 L_{\mathrm{it}} + a_3 \mathrm{BB\_Pen}_{\mathrm{it}} + \varepsilon_{1\mathrm{it}} \tag{11.7}$$

Demand function:

$$\mathrm{BB\_Pen}_{\mathrm{it}} = b_1 \mathrm{BBPr}_{\mathrm{it}} + b_2 \mathrm{GDPC}_{\mathrm{it}} + b_3 \mathrm{Edu}_{\mathrm{it}} + b_4 \mathrm{Urb}_{\mathrm{it}} + \varepsilon_{2\mathrm{it}} \tag{11.8}$$

Supply function:

$$\mathrm{BB\_Rev}_{\mathrm{it}} = c_1 \mathrm{BBPr}_{\mathrm{it}} + c_2 \mathrm{GDPC}_{\mathrm{it}} + c_3 \mathrm{HHI}_{\mathrm{it}} + \varepsilon_{3\mathrm{it}} \tag{11.9}$$

Output function:

$$\Delta \mathrm{BB\_Pen}_{\mathrm{it}} = d_1 \mathrm{BB\_Rev}_{\mathrm{it}} + \varepsilon_{4\mathrm{it}} \tag{11.10}$$

According to the model, mobile broadband appears to have an initial effect on the economy. Contrary to the 2010 results, every 1 % increase in mobile broadband penetration yields 0.022 % growth in GDP (see Table 11.8).

Katz, R. (2010). The contribution of broadband to economic development. In V. Jordan, H. Galperin, & W. Peres (Eds.), *Fast-tracking the digital revolution: Broadband for Latin America and the Caribbean*. Santiago: UN Economic Commission for Latin America.

Katz, R. (2011). *"The impact of broadband on the economy: research to date and policy issues", Trends in Telecommunication reform 2010-11*. Geneva: International Telecommunication Union.

Katz, R. (2012). *The impact of broadband on the economy: Research to date and policy issues* (The impact of broadband on the economy broadband series). Geneva: International telecommunication Union.

Katz, R. L., Avila, J., & Meille, G. (2010a). *Economic impact of wireless broadband in rural America*. Washington, DC: Rural Cellular Association.

Katz, R., Avila, J., & Meille, G. (2011). *The impact of wireless broadband in rural America*. Washington, DC: Rural Cellular Association.

Katz, R., & Callorda, F. (2011). *Medicion de Impacto del Plan Vive Digital en Colombia y de la Masificacion de Internet en la Estrategia de Gobierno en Linea*. Bogota: Cintel. Diciembre 2.

Katz, R., & Suter, S. (2009). *Estimating the economic impact of the broadband stimulus plan*. Columbia Institute for Tele-Information Working Paper. Retrieved July 28, 2010 from www.elinoam.com/raulkatz/Dr_Raul_Katz_-_BB_Stimulus_Working_Paper.pdf

Katz, R., Vaterlaus, S., Zenhäusern, P., & Suter, S. (2010b). The impact of broadband on jobs and the German economy. *Intereconomics, 45*(1), 26–34.

Koutroumpis, P. (2009). The economic impact of broadband on growth: A simultaneous approach. *Telecommunications Policy, 33*, 471–485.

Waverman, L. (2009). *Economic impact of broadband: An empirical study*. London: LECG. February 29.

Chapter 12
Africa Leapfrogs into the World Economy

Melissa T. Cook

African Sunrise Partners is dedicated to bringing private sector capital into Africa—for growth and profit. We provide information on business and investment opportunities and work to clarify misperceptions about the current situation in key African countries, using insights gained during regular trips to Africa. We've had many conversations with otherwise worldly and sophisticated investors, corporate executives, lawyers, and accountants who still think of Africa as a place of war, corruption, and poverty. In a rising number of African countries—home to hundreds of millions of people—the story is very different. During our visits to the continent, we see a vibrant, dynamic Africa filled with entrepreneurs building businesses. The continent is sometimes noisy and chaotic but is also very much worth a closer look for investors.

Why is Africa changing so rapidly, and why now after so many years of failed promise? We think communications and technology are playing a major role. Costs are falling while capabilities soar for information and communications technology (ICT) at the enterprise and the consumer level. Africa's tech-savvy young population is ready, willing, and able to adopt new ways of communicating and doing business. Voters with information and current news are active participants in determining the direction their country will take. This is a powerful combination for business, democracy, and broad-based economic growth.

In this chapter, we outline some of the background themes to our "Why Africa, Why Now?" story. We analyze the business and financing elements of the communications and broadband video story and provide perspective on the types of companies we believe will play a role in the industry's growth. And for the benefit of readers outside the corporate or the investment communities, we offer some thoughts on how these groups make their investment decisions and how this is relevant to the broadband video story in Africa.

M.T. Cook (✉)
African Sunrise Partners LLC, Patterson, NY, USA
e-mail: cook@afrsun.com

J. O'Neill et al. (eds.), *Broadband as a Video Platform*, The Economics of Information, Communication, and Entertainment, DOI 10.1007/978-3-319-03617-5_12,

Why Africa, Why Now?

Africa is on the move—right now. The days of Africa being seen as the "hopeless continent" are in the past. Africa is not without its problems. However, we believe that the time is right for investors and companies to take a serious look at how they plan to invest and build a profitable, competitive business on the continent. It's not too early—markets are already becoming well developed and long-term service relationships are forming, while buyer preferences clearly favor ultramodern technology at an accessible price point.

We see three broad growth themes in business in Africa that we believe should be top of mind for the management teams and directors of international companies—and for their shareholders: Governance is improving, money is pouring into infrastructure, and the emerging middle class is quickly developing a taste for branded goods and quality services.

- *Improving governance: universal and open access to information and ideas is essential to this story.* Risk goes down when countries run free and fair elections between viable political challengers. Each time an election yields a peaceful handover of power from one political leader to another, it raises the odds that the next election on the continent will proceed peacefully and represent the will of the people. During the past few years, we have seen elections yield a peaceful political transition—even if in a messy or an imperfect fashion—in countries including Zambia, Senegal, Kenya, Ghana, and Nigeria. New constitutions and more consistent enforcement of existing laws can produce greater certainty and stability in a country's economy, which comes as welcome news to investors.
- *Urbanization and rise of a middle class*: Africa is urbanizing rapidly. People move into cities in search of better jobs, services, and access to education. Cities expand to incorporate rural or suburban areas. Urban dwellers generally have higher incomes, and they spend more on goods and services. Consumer spending is on the rise. Even the poorest shoppers have money to spend. Thriving informal economies provide spending power not captured in official statistics—creating underappreciated levels of demand. Formal markets cater to middle-class shoppers who flock to gleaming new malls offering a dizzying array of choices. Information and communication services are a vital part of the budgets of urban consumers.
- *Infrastructure to enable industrialization*: The Chinese proverb holds true in Africa: "To get rich, first build a road." All over Africa, new roads are being paved. Rail lines are being rebuilt, airports are being expanded, and ports are being upgraded. Electricity is still too scarce and too expensive, but there are encouraging signs as key countries privatize the sector in order to attract investment in new capacity. Alternative energy installations are becoming viable thanks to shrinking costs for solar and other technologies. Expanded infrastructure enables industrialization and supports economic growth. China's engagement has yielded the rapid rollout of new infrastructure capacity. Supply begets

demand—more telecom, road, rail, power, airport, and port capacity is a clear stimulus to economic growth and demand.

Broadband Video: An Enabler and a Beneficiary of These Trends

ICT is an essential element of our positive case for Africa. Counterintuitive though it may sound, Africa is well positioned today to benefit from new technology specifically because the continent lacks legacy systems and entrenched, inflexible competitors and regulatory infrastructure.

Africa has a shortage of bricks-and-mortar banking and retail outlets, libraries, efficient government offices, educational and healthcare facilities, and agricultural extension offices. It's expensive to manage a business, run a household, or provide government services without the benefit of a developed information, transportation, and financial services infrastructure. Shopping, registering a car, transferring money, and paying bills can be time consuming and usually involve carrying large amounts of cash. Commercial and agricultural markets are inefficient because buyers and sellers don't have access to current market-based prices.

Technology is addressing many of these challenges. Mobile money, videoconferencing or data transmission for healthcare, and online education all are viable new businesses. We see a tremendous amount of innovation happening in Africa—in ICT and other sectors. Local entrepreneurs are designing solutions to local problems. They are supported in many areas by major multinationals. An IBM, Google, or Microsoft venture, R&D effort, or data center may directly or indirectly underpin the development of local developers, service providers, and businesses. Multinationals train and develop experts who often go on to build their own local ventures, creating valuable local jobs and developing new intellectual property that will be used in Africa and throughout the world.

Communications Investment in Africa Is Nothing New

The African communications sector has already received investment that caused nothing short of a revolution. The sector's influence on economic performance has already been significant. Investments made and services developed during the past decade are the foundation for the next generation of communication advances on the continent.

What were the catalysts? What was the result? Consider:

- *Vision*: In the late 1990s, Sudanese entrepreneur Mo Ibrahim saw the demand for mobile phone services that others overlooked because of their reliance on statistical evidence rather than actual market knowledge. Rather than seeing the low levels of fixed-line penetration as indicative of poor demand, Ibrahim believed that properly priced mobile services would be taken up by millions upon tens of millions of consumers. His flexible, affordable mobile model gained near-universal acceptance, created an entirely new industry in Africa, and gave others confidence to invest in mass-market mobile services that have been wildly successful.

- *Competition*: Enlightened regulators in key markets recognized that open and free competition amongst private sector mobile operators would drive prices down and accessibility of mobile services up. Technology suppliers developed the capability to offer prepaid airtime, sold by vendors in informal and formal markets. Africa benefited from the entry of aggressive Chinese equipment manufacturers whose technology was priced below that of global competitors but offered sufficient functionality to meet Africa's needs.
- *Consumer demand*: Africans quickly demonstrated an insatiable appetite for communications. Our favorite illustration of this is a billboard we saw in Lagos, which freely acknowledges the propensity of Africans to talk, talk, talk—and offers them airtime at a low, low price (Figs. 12.1 and 12.2). Figure 12.3 illustrates the rapid expansion in mobile penetration in major countries over the past 10 years, while Fig. 12.4 illustrates the total mobile phone users on the continent. We often ask Africans about consumer spending priorities, with the question, "What would you do with an extra 100 shillings, 5 cedi, or 50 naira?" The answer is invariably "Buy airtime." Affordable, ubiquitous mobile phone service allows people to stay connected, learn about employment opportunities, and share information.
- *Declining prices*: Ultra-cheap "feature phones" (e.g., basic mobile phones with no smartphone features) and inexpensive packs offering prepaid voice minutes allow entry-level consumers to get connected. In the most competitive markets, operators have cut airtime tariffs aggressively to stimulate demand. Consistent with the historical pattern in other markets, average revenue per user declines as penetration extends to include lower intensity users. Demand for minutes is not as elastic as in developed markets, primarily because of low consumer spending power in large segments of the market. Sometimes, price cuts just mean lower revenue, not greater volume.
- *Expanding services*: In the longer term, we see data and other advanced services as key drivers of the sector's revenue growth. Figure 12.5 illustrates Cisco's estimates for global data growth—and Africa's rising contribution to the total. For mobile operators, rising demand for data delivery offers an opportunity for improved revenue and margin performance.
- *Favorable demographics*: Demographics work in Africa's favor: Younger Africans are eager to adopt new technologies. Some of the most exciting and dynamic people we've met on the continent are young graduates who see the opportunity to solve problems, make money, and blaze a new path in the ICT sector. Social media is thriving across the continent.
- *Scale/innovation*: A multi-hundred-million-strong user base offers scale to justify the development of new communication-related services. Entrepreneurs are creating locally designed solutions to meet African needs. The most visible is

Fig. 12.1 Talk is cheap. *Source*: African Sunrise Partners LLC

Fig. 12.2 Prepaid airtime cards. *Source*: African Sunrise Partners LLC

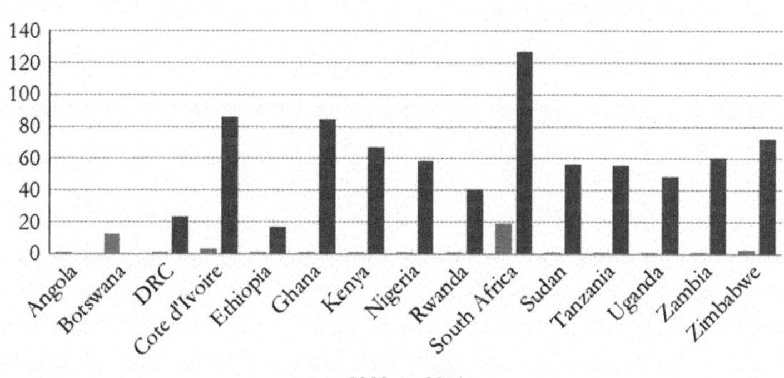

Fig. 12.3 Mobile penetration rate—2000 vs. 2011 (ITU). *Source*: International Telecommunication Union (ITU)

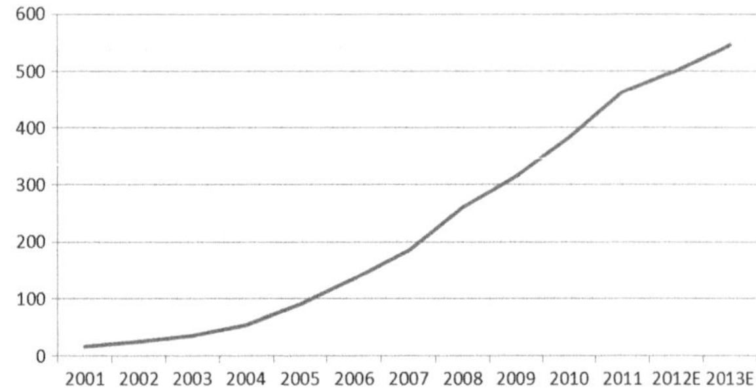

Fig. 12.4 Mobile cellular subscribers in Africa, in millions (ITU). *Source*: International Telecommunication Union (ITU)

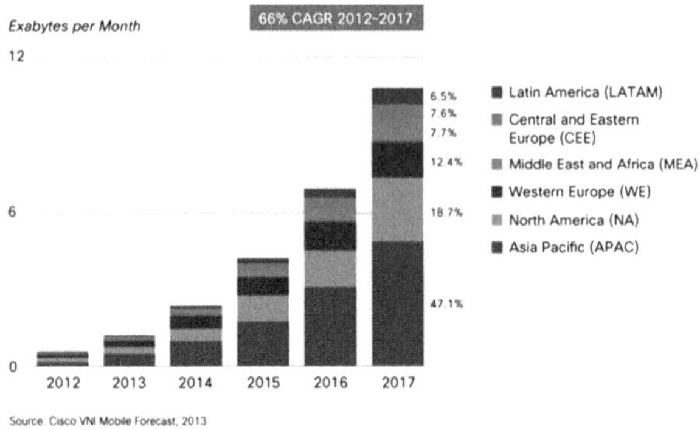

Fig. 12.5 Projected mobile data traffic growth (Cisco). *Source*: Cisco (*Visual Networking Index: Global Mobile Data Traffic Forecast Update, 2010–2015*, here)

Kenya's mobile money platform M-Pesa, which has brought millions of new customers into the financial system. It has facilitated transactions of all sizes between people throughout the country, boosting the velocity of money and adding measurably to Kenya's economic growth. Technology used in the Ethiopia Commodities Exchange allows rural farmers to access larger markets and benefit from price transparency. Kenya's iCow brings text-based information to dairy farmers, helping them optimize milk yields and boost income.

Networks to Support Soaring Demand

Africa is no longer a continent separated from major economies by silence. Subsea cables offering vastly expanded data capacity landed in key markets including Kenya, South Africa, and West Africa during the past several years. Africa is fully connected to the world with bandwidth transmitting ever-expanding quantities of data. The carrying capacity of these cables has risen exponentially based on rapid advances in the electronics controlling data running through the system. The map in Fig. 12.6 highlights the subsea cable capacity already in use and slated to come online over the next several years. Sub-Saharan Africa has already seen soaring private sector infrastructure investment in communication assets (Fig. 12.7).

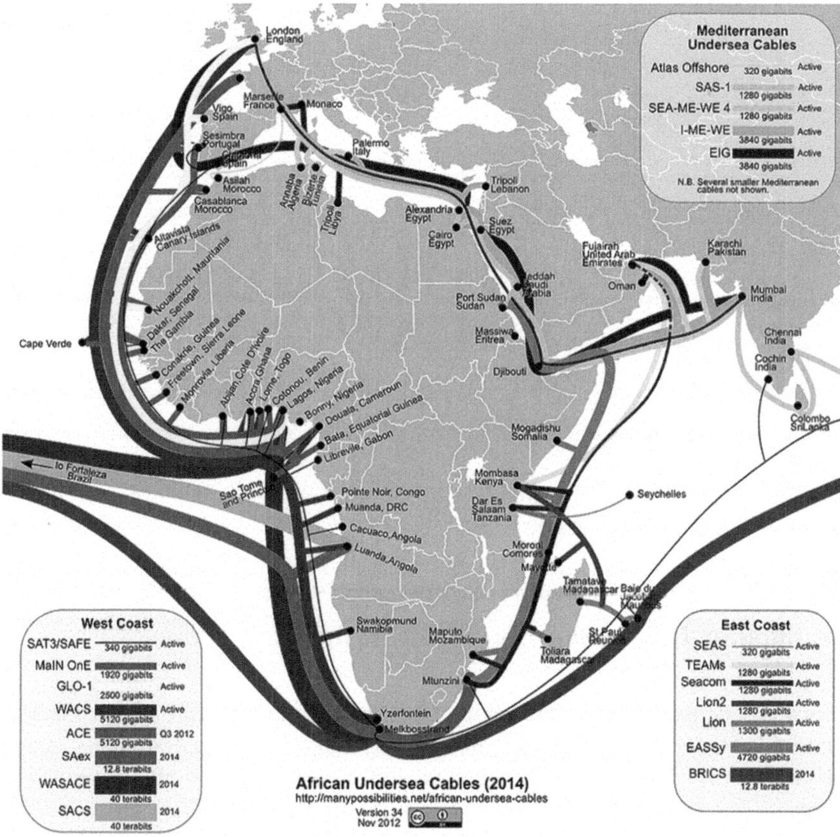

Fig. 12.6 Subsea cables bring data transmission capacity to Africa (Many Possibilities). *Source*: Printed courtesy of Steve Song, Many Possibilities (www.manypossibilities.net). License: Creative Commons Attribution CC BY 3.0

Bringing data to Africa isn't enough, however. Companies are now tackling bottlenecks in the system from the fiber landing sites through to the last mile into customer premises. Data must be distributed via wireless or fixed-line networks. Bandwidth is being pushed further out into the network, via metropolitan fiber rings, long-haul fiber runs, mobile, microwave, or satellite channels. One illustration: Liquid Telecom is building terrestrial fiber networks in southern and eastern Africa (Fig. 12.8), sometimes in cooperation with electric utilities for broadband over power lines. This process isn't always smooth. Service disruptions are common across the continent because of "backhoe fade" (construction-related fiber cuts) or cable and equipment theft.

Companies are putting up new microwave transmission capacity in areas where fiber links are too expensive. Operators are investing in advanced systems for efficient mobile spectrum use or new towers and building local last-mile fiber or Wi-Fi networks in urban areas. Urban airwaves are clogged with heavy voice and

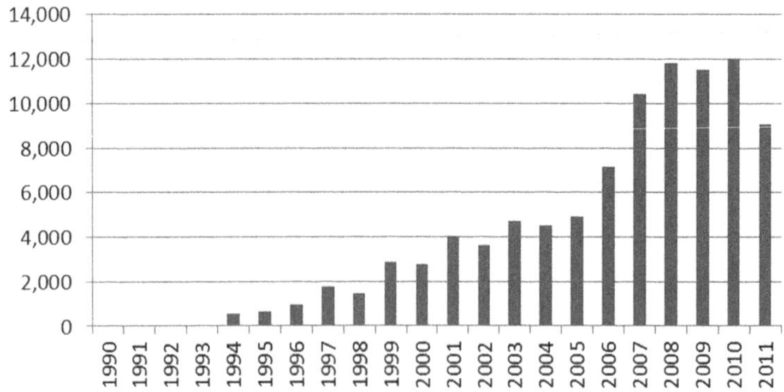

Fig. 12.7 Total private investment in sub-Saharan Africa telecom (US$, mn; World Bank). *Source*: 2011. *PPI Database*. ©World Bank. License: Creative Commons Attribution CC BY 3.0

data traffic. Governments need to get to work quickly on allocating new spectrum and allowing competition to invest and flourish where more capacity is needed.

Technology Innovation Happening at the Right Time

Africa is benefiting from an innovation "race to the bottom" in the mobile handset space. Encouraged by soaring demand from developing-market consumers, Asian electronic firms are developing equipment offering ever more features at ever-lower prices.

While the latest iPhone gets all the attention in developed markets, in Africa the Holy Grail is the sub-$50 smartphone. This is the price at which mass adoption is expected to happen, pushing mobile Internet usage dramatically higher. Samsung, Huawei, HTC, and ZTE are some bold-faced names in this space. Microsoft's recently announced venture with Huawei for Windows-enabled, low-priced phones is a further illustration of this story. Household computer ownership is quite low, so we expect many Africans to use mobile devices to access data and video services. Of course, deployment of smartphones will put even greater pressure on urban airwaves, underpinning the demand for mobile network equipment.

Cloud computing is becoming the norm for global business. Removing the need for bulky, expensive, power-hungry, and maintenance-intensive in-house enterprise hardware allows African entrepreneurs to access and afford the services they need to run their companies. Cheaper and more accessible centralized computing capacity will allow developers to build transformative services for government, education, healthcare, financial services, and industrial users.

Data centers are in short supply in Africa so far because of costly land, connectivity, and power requirements. However, electricity generation capacity is improving and will eventually lead to reduced costs for power users throughout the industry. Declining costs for distributed (off-grid) or alternative energy sources

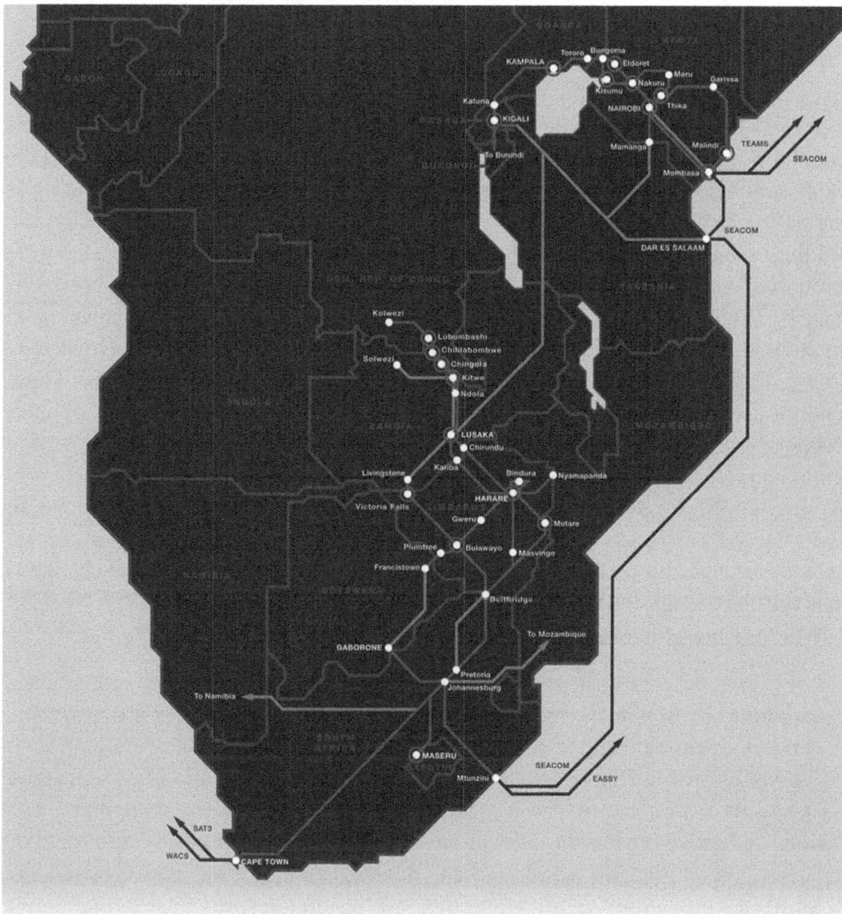

Fig. 12.8 Liquid Telecom is building terrestrial fiber networks in southern Africa (Liquid Telecom). *Source*: Liquid Telecom

and deployment of long-haul fiber capacity will reduce barriers to entry here. Suppliers of servers, data storage devices, routers, and electrical and cooling equipment need to focus now on addressing the requirements of African data center operators. We have seen complaints about designs leading to unnecessarily high power consumption in developed-world data centers, where power generation capacity is plentiful and conservation isn't a high priority. In Africa, data center operators will demand the most modern, efficient technology at an accessible price point.

Bringing Broadband Ideas to Reality: Technology and Economics

Investors thinking about how to invest in Africa's burgeoning communications sector will find activity in a variety of areas. But the pattern of development and the list of suppliers may look very different than in developed markets. Innovation is originating from new and different places: Developing-market entrepreneurs are building systems, equipment, and solutions that are appropriate for emerging markets. Competition from Chinese, Korean, and Indian equipment suppliers is intense. Firms based in developing markets often operate from a lower cost structure than firms selling to richer customers. Developing-markets firms may have a different design sensibility than entrenched global competitors. These vendors also have a laser-like focus on the price/performance balance of their products. Across industrial and technology sectors ranging from construction machinery to ultrasound devices to consumer electronics, we see products that are engineered to deliver adequate performance at price levels acceptable to emerging-market buyers, rather than being engineered to perfection and at a price level to match. Tables 12.1 and 12.2 illustrate some of the companies and segments involved.

Where is investment money being spent? Here are a few examples:

- *Laying the pipes*: Construction firms, fiber and cable manufacturers, engineers, and network architects are already at work. Since fiber cuts are common, networks must be engineered with redundant transmission routes.
- *Putting up the towers*: Mobile phone towers are pushing out into very rural areas. Desperately overloaded urban systems require more towers, more fiber, and more effective spectrum allocation. Towers require primary and backup power, involving grid power (where this is available) and diesel generators (protected by a strong security system and a full-time security guard). Innovative systems to use alternative energy and hybrid power generation are likely to find strong demand as prices decline. Examples of technology being developed to improve the performance of towers include colocation of transmission equipment from more operators, stronger signal broadcast with reduced power consumption, and improved quality of signal transmission for equipment sited lower on a tower.
- *Lighting the fiber*: Laying fiber is one part of the story, and lighting it is another. Capacity doesn't exist until electronics—and the power and software to run the network and route the data—are in place.
- *Installing customer premise equipment, software, and services*: Asian OEMs including Lenovo and Samsung are very active and visible in Africa. IBM HP, Microsoft, Cisco, SAP, and Oracle are pushing for a larger share of the enterprise hardware, software, and services markets. Google is actively engaged in developing e-commerce and other solutions to African needs, many of which will have utility in other global markets.

Table 12.1 Companies of all stripes involved in African ICT (ASP)

Company	Business activities
Access Kenya	Internet services
Alink Telecom	IP-based telecommunication solutions for Africa
AMD Global Telemedicine	Remote medicine applications
Asia Broadcast Satellite	Satellite coverage of portions of sub-Saharan Africa
AT&T	Global telecom services
Bharti Airtel	Mobile communication services
Businesscom Networks	Satellite Internet solutions covering sub-Saharan Africa
Canal + (Vivendi)	Satellite-based programming covering sub-Saharan Africa
Cisco	Networking and telecom equipment
DUMA	Mobile-based job-matching service
Econet	Telecommunication services
Econet Wireless	Mobile communication services
Emerson	Network power, data center equipment, other electrical equipment
Ericsson	Telecom and data networking equipment
Essar Telecom	Mobile communication services
Eutelsat Communications	Satellite coverage of portions of sub-Saharan Africa
Google	Internet services
Helios Towers	Mobile tower operator
Hewlett-Packard	Hardware, software, and services
Huawei	Telecom and data networking equipment
IHS	Mobile tower operator
IBM	Hardware, software, and services
iCow	SMS-based information service for dairy farmers
Jamii Telecoms Ltd.	Network infrastructure services
Kenya Data Networks	Internet and network services
Lamit Company	Satellite coverage of portions of sub-Saharan Africa
Lenovo	Computer technology for developing-market users
Lion	Subsea cable operator
Liquid Telecom	Terrestrial fiber networks and services
Main One	Subsea cable operator
Microsoft	Software and services
Mobitel	Internet services
Motorola	Hardware and services
Hitachi Data Systems	Storage and cloud computing equipment
MTN	Mobile communication services
Nokia	Telecom and data networking equipment
Oracle	Software and services
Orange	Mobile communication services
Safaricom	Mobile communication services
Samsung Electronics	Hardware provider
SAP	Software and services
SasaAfrica	E-commerce platform connecting users via mobile phone
Seacom	Subsea cable operator
SES	Satellite coverage of portions of sub-Saharan Africa
Speedcast	Satellite communication services
Suburban Telecom	Terrestrial fiber network operator

(continued)

Table 12.1 (continued)

Company	Business activities
TEAMs	Subsea cable operator
Transition Networks Inc.	Hardware manufacturing
Vizada	Satellite and value-added communication services
Vodacom	Mobile communication services
WIOCC/Eassy	Subsea cable and network services
Zain	Mobile communication services
ZTE	Telecom and data networking equipment

Source: African Sunrise Partners LLC

- *Keeping it local*: Bold-faced names of multinational companies tell only part of the story, however. Africa has a vibrant ecosystem in IT innovation. Governments including Kenya and Nigeria are investing to facilitate collaboration and support the development of new services; but in our view the most exciting action is taking place in the private sector, often at the micro-enterprise level.

What's going to be pushed through all of these networks? Who will pay?

This entire story has a bit of a chicken-and-egg aspect to it. Network construction must be funded by revenues from users. Financial forecasting of revenues requires an understanding of how people will use networks, what they will pay for, and how all of this investment will generate returns. The runaway success of Mo Ibrahim's mobile business—at a time when there was no statistical evidence that people would pay for communication services—illustrates the challenge here. On one hand, content developers and network operators must believe that there will be demand for more content than is being consumed today. On the other hand, you can't build a bankable business plan on hope and dreams alone. Our optimism is based on the evidence that broadband networks can deliver far more than just entertainment and information. Innovative entrepreneurs can use available bandwidth to transform how people interact with business, government, healthcare system, and schools.

Here is a sampling of content development initiatives we've seen:

- *Mobile banking and financial services*: Kenya's M-Pesa is the best known, but mobile money is an idea whose time has come continent-wide. From receipt of remittances to cross-border payment of business accounts to efficient payment of school fees and utility bills, mobile money is everywhere. As middle-class Africans start buying insurance, accessing lines of credit, and opening retirement savings accounts, this growth will not be hampered by the limited bricks-and-mortar banking presence across the continent.
- *E-government*: From voter registration, property title transfer, and digital access to court records to a public means of tracking how tax money is being spent, investments in e-government are likely to yield quick returns to citizens.
- *Healthcare*: Examples here include transmission of digital diagnostic test results to a remote physician and videoconferencing between patients in rural clinics and urban medical specialists.

Table 12.2 Many layers of investment required to deliver broadband video (ASP)

Market segment	Selected market participants
Submarine cable systems	• EASSy • MainOne • Seacom • TEAMs • WACS • WASACE • Corning
Equipment manufacturers	• Ericsson • Cisco • Huawei • ZTE
Data center technology and services	• Google • Eaton • Emerson • IBM • Microsoft
Consumer equipment	• Motorola • Nokia • Samsung • HTC • Lenovo
Mobile tower operators	• American Towers • Helios • IHS
Broadband along power lines	• Copperbelt Energy • Liquid Telecom
Power	• Diesel generators (e.g., Cummins, Caterpillar) • Solar • Hybrid alternatives
Network operators	• Liquid Telecom • Airtel • MTN • AT&T • Incumbent wireline providers
Financial institutions and payment platforms	• M-Pesa • Airtel Money • Google's Beba NFC payment card • Virtual Terminal Networks
Innovators, disruptors, and content providers	• iCow • iSchool.zm • Silicon Savannah • Nollywood • Google • Paga • m-Health, e-Learning

Source: African Sunrise Partners LLC

- *Education*: Access to remote libraries, data, and information is being enabled by technology. Students living far from cities can now access specialist educators and higher level teachers via video links.
- *Commerce*: Africa's millions of small-scale entrepreneurs are able to tap into far-flung markets, sources of materials, pricing data, and financial support.
- *Local news and content*: Given the extreme fragmentation of African markets, this content needs to be of low cost and high impact. However, it can also bring people together to share how problems were solved in a neighboring country, how societies are evolving, and how African regions can unite to create a stronger economy.
- *Entertainment*: Nigeria's Nollywood is already world famous for its film production, which is the classic example of pricing a product to meet the needs of its buyers. Movie "theaters" in Africa today can be as simple as a room with chairs and a white wall; when the market will bear higher prices, they can be as plush and high-tech as people desire.

What's the Financial Angle?

How can African governments attract investor funding? They must start by involving the private sector as a partner in designing a stable and predictable regulatory framework. This will allow companies and financiers to assess markets, develop viable long-term business plans, allocate capital, and build successful business models.

Government policymakers need to understand how corporate executives think. Corporate managements are stewards of shareholder capital, and they answer to a variety of stakeholders. The chief executive officer, chief financial officer, board of directors, strategic planning department, and operating management all are focused on allocating capital in a way that generates returns on capital in excess of the cost of capital. If this math doesn't work, an investment or an acquisition will destroy value. Destruction of value is anathema to the purpose at the heart of even the most altruistic, socially minded companies—i.e., to create value for stakeholders. No corporation lives with a value-corroding element in its midst for very long—unless deemed warranted by a solid business case anchored to visible, quantifiable future growth potential. Firms with a public listing on a US stock exchange also face relentless pressure from investors to deliver consistent quarterly earning growth, a goal that often is incompatible with long-term business development in markets of uncertain size. Asian firms tend to think in terms of decades, not quarters, and they often take a longer term, more strategic approach to capital allocation.

We see rising demand for investments in Africa from institutional investors. These organizations are responsible for delivering returns to their clients. These clients include college endowments, pension plans, and other savings vehicles that must yield a return to their owners. Institutional investors plan their asset allocation and investment programs to deliver a particular combination of risk-adjusted returns. Africa can provide an important element of diversification and growth potential to a portfolio, when viewed in the context of the balance of that investor's holdings.

Investing in Africa poses opportunities and challenges for companies, investors, and entrepreneurs:

- *Perception of challenges is often greater than the reality.* We see a general lack of awareness and poor understanding of opportunities in Africa. Global investors and corporations are held back by spotty access to information about Africa. News in mainstream developed-market outlets is often distorted, with a focus on shock value rather than balance. In this environment, risks are magnified while opportunities are underplayed. Real risks that we see include high costs because of poor infrastructure and limited human capital, realities of disease and difficult physical environments, changing regulations, and uneven application of rule of law in different countries. Corruption is a factor in Africa, but it is not unique to the continent.
- *The opportunities are compelling, in our view.* Demand exceeds supply for virtually everything. Spending power is far greater than is indicated by official statistics, which fail to capture the size of informal markets. Many items have latent demand: How can you measure electricity or cement consumption when supply isn't there? Companies often find that demand far exceeds their original estimates, forcing them to scramble to expand capacity far earlier than market data had indicated.
- *Unique time and place for entrepreneurs and vendors in the ICT sector.* We see a particularly exciting opportunity in the ICT sector. Entrants have an opportunity to start from scratch, without the need to work around shortcomings of legacy systems. Networks and systems can be designed to avoid the costly mistakes made in developed markets where architecture is now destiny. Companies can design networks with the benefit of cheaper technology, delivering increasingly robust performance. Network engineers have a far clearer idea of how voice, video, and data traffic flow than was possible in recent decades when developed markets' cable TV networks were built. Best of all, Africa's innovators are coming into their own, building solutions to their continent's problems and positively affecting how Africa develops in coming decades. This is the dynamic marketplace, rich with potential, that today's new entrants into African ICT markets step into.

Contributor Biographies

Ben Akoh is a researcher and expert in education, ICTs, media, Internet, and technology policy. His research interests are in exploring the nexus of education, culture, and the Internet for sustainable growth and development. He is an instructor at the University of Manitoba delivering technology-based courses on emerging technologies, open education resources, and digital literacy and has supported various pro-democracy and open dialogues in media and Internet initiatives in Africa and globally. He works for a policy think tank in Canada, with past private, non-profit, and UN experiences. He has lived and worked in East, West, and Southern Africa.

Brian Arendse started his career as a Radar Systems Engineer at the Council for Scientific and Industrial Research (CSIR) in Pretoria, South Africa. He has more than 15 years industry experience. In Ericsson, Brian has fulfilled R&D, Network Design, Solutions Management, Sales, Marketing, and Strategy roles in Sweden, South Africa, and Nigeria. Brian has a B.Sc. in Electrical Engineering from the University of Cape Town, South Africa, an M.Sc. in Electrical Engineering from Chalmers University of Technology, Gothenburg, Sweden, and is completing an M.B.A., at Wits University, South Africa.

J. O'Neill et al. (eds.), *Broadband as a Video Platform*, The Economics of Information, Communication, and Entertainment, DOI 10.1007/978-3-319-03617-5,

Melissa T. Cook, CFA, is the Founder and Managing Director of African Sunrise Partners LLC. ASP is an investment strategy firm dedicated to bringing private sector capital into sub-Saharan Africa. Visit www.afrsun.com for more information.

Ms. Cook has worked in the global equity business for over 25 years. She covered entertainment, broadcasting, housing, capital goods, and consumer products companies at Drexel Burnham Lambert, Prudential Securities, and CLSA/Crédit Agricole. At CLSA, Ms. Cook was the Head of Asia USA Strategy, where she visited China regularly and wrote about the impact of China's rise on multinational companies and global portfolios. She also worked as Head of Global Research at Lazard Asset Management.

In her work on sub-Saharan Africa, Ms. Cook draws on her experience in Central and Eastern Europe during the years after the fall of Communism and her knowledge of how China's economy and global ambitions are evolving. For her research on broadband video in Africa, Ms. Cook used knowledge gained during her years covering the US and UK cable television sectors as they rolled out fiber networks and developed advanced digital services.

Ms. Cook has an A.B. degree in History from Dartmouth College and an M.B.A. in Finance from the Stern School of Business at New York University. She is a CFA charterholder and holds FINRA Series 7, 16, 24, 63, 86, and 87 licenses.

Indra de Lanerolle is an experienced communicator, teacher, and strategist who has worked with major organizations in South Africa and internationally, including AVUSA, BBC, GAVI Alliance, Monitor Group SA, Multichoice, SABC, Soul City Institute for Health and Development Communication, South African Department of Health, Telkom, and Unilever SA.

He is an award-winning film and television producer with 20 years experience working in over 15 countries in Europe, Africa, the United States, and Asia.

A former media entrepreneur and company director, he is an adjunct lecturer at the University of Witwatersrand Journalism and Media Programme and is lead researcher of the South African Network Society Survey. He is a regular speaker and commentator on the impact of the new network age on organizations, media, social development, and the economy.

He lives in Johannesburg, South Africa.

Darcy Gerbarg is currently the Director of Operations for CineGrid, Inc., an international technology nonprofit, whose members do experiments with very large media files over the Lambda Rail. She is a Senior Fellow at the Columbia Institute for Tele Information (CITI), Columbia University Business School, since 1997. In the past, she was the Executive Director of the Guglielmo Marconi International Fellowship Foundation, Inc. (the Marconi Society, Inc.) at Columbia University. Prior to that she held a research position at Courant Institute for Mathematical Sciences, New York University and as an adjunct professor has taught at many universities. Ms. Gerbarg is a pioneer computer artist. Her work has been exhibited in galleries and museums internationally.

Ms. Gerbarg is the series editor for *The Economics of Information, Communication and Entertainment: The Impacts of Digital Technology in the 21st Century*, Springer.

Alison Gillwald (Ph.D.) is Executive Director of Research ICT Africa, an 18-African-country ICT policy and regulatory research network based in Cape Town, South Africa, which is funded to support evidence-based policy development on the continent. She is Adjunct Professor at the Management of Infrastructure Reform and Regulation Programme at the University of Cape Town Graduate School of Business. She served on the founding Council of the South African Telecommunications Regulatory Authority (SATRA) and prior to that established the Policy Department at the first Independent Broadcasting Authority. She has served on several public boards and advisory bodies including chairing the National Digital Broadcasting Advisory Body. She has consulted widely for multilateral agencies and governments and has published in the areas of telecommunications and broadcasting policy and regulation, gender, and politics more broadly.

Dr. Heather E. Hudson is Director of the Institute of Social and Economic Research (ISER) and Professor of Public Policy at the University of Alaska Anchorage. Previously, she was Founding Director of the Communications Technology Management and Policy Program at the University of San Francisco. Her work focuses on applications of ICTs for socioeconomic development, regulation, and policy issues.

Prof. Hudson has planned and evaluated communication projects in Alaska, northern Canada, and more than 50 developing countries and emerging economies in Africa, Asia, Latin America, the Caribbean, the Middle East, Eastern Europe, and the South Pacific. She has consulted for the private sector, government agencies, consumer and indigenous organizations, and international organizations.

She has written many articles and several books and has presented numerous conference papers as well as expert testimony on communications policy issues such as universal service and access, incentives for investing in information infrastructure, restructuring of the telecommunications sector, and telecommunications planning for socioeconomic development.

Prof. Hudson has been a board member of the Pacific Telecommunications Council, the Telecommunications Policy Research Conference, Women in Telecommunications (WiT), Farm Radio International, and the International Council for Computer Communications (ICCC). She serves on the editorial boards of Telecommunications Policy, Information Technologies and International Development, and the Journal of Community Informatics. She has been a member of Advisory Committees of the US National Research Council, the Federal Communications Commission, the Department of Commerce, and the Office of Technology Assessment.

Dr. Raul Katz joined CITI in April 2007 as the Director of Business Strategy Research. He is currently an Adjunct Professor at Columbia Business School, where he teaches a Seminar in International High Technology Strategy within the M.B.A. program. Dr. Katz has been a management consultant in the telecommunications industry for the past 25 years. After 20 years of service, he retired as a Lead Partner and Head of the Telecommunications Practice in the Americas with Booz Allen Hamilton. He then served as CEO of Adventis, an international telecommunications consulting firm. He also serves as President of Telecom Advisory Services, a strategy consulting firm. Dr. Katz has published extensively in research journals; his books include "The Information Society: an International Perspective" and "Creative Destruction: Business Survival Strategies in the Global Internet Economy." Dr. Katz has a Ph.D. in Political Science and Management and an M.S. in Communications Technology and Policy, both from MIT, as well as a Maitrise in Political Science and a Maitrise in Communication Sciences from the Sorbonne.

Dr. Tim Kelly is a Lead ICT Policy Specialist in the ICT Sector Unit of the World Bank in Washington DC. He leads the analytical work program on ICT for Development, which is shared with *info*Dev. He has been task team leader on the *Creating Sustainable Businesses for the Knowledge Economy* and *eTransform Africa* programs. He was formerly Head of the Strategy and Policy Unit of the International Telecommunication Union (ITU) and has previously worked with the OECD and Logica Consultancy Ltd.

Over the last 25 years, Dr Kelly has specialized in the economics of information and communication technologies. He has written or coauthored more than 30 books on the subject including the World Bank's "*Building Broadband*," ITU's "*Internet Reports*" and "*World Telecommunication Development Report*," and OECD's "*Communications Outlook*." He has an M.A. (Hons) degree in geography and a Ph.D. in industrial economics from Cambridge University.

Dr. Koutroumpis is a Research Fellow at the Innovation and Entrepreneurship Group of Imperial College Business School and a Fellow of the Columbia Institute of Tele-Information at Columbia Business School. He has worked on projects with the World Bank, the European Investment Bank, the International Telecommunications Union, the OECD, major telecommunications' operators, and equipment vendors. He received the 2011 Dissertation Award by the Telecommunications and Public Utilities Group of the American Economic Association and the 2010 Principal's Award for the "Most Outstanding Doctoral Thesis" by Imperial College London. He received his Ph.D. in Economics from Imperial College London and an M.Phil. in Technology Policy from Cambridge University.

Dr. Eli Noam has been Professor of Economics and Finance at the Columbia Business School since 1976 and was recently named the Garrett Professor of Public Policy and Business Responsibility. In 1990, after having served for 3 years as Commissioner with the New York State Public Service Commission, he returned to Columbia. Noam is the Director of CITI. He also served on the White House's President's IT Advisory Council. Besides the over 400 articles in economics, legal, communications, and other journals that Professor Noam has written on subjects such as communications, information, public choice, public finance, and general regulation, he has also authored, edited, and co-edited 28 books. Noam has served on the editorial boards of Columbia University Press as well as of a dozen academic journals and on corporate and nonprofit boards. He is a regular columnist on the new economy for the Financial Times online. He is a member of the Council for Foreign Relations and a fellow of the World Economic Forum. He received AB, AM, PhD (Economics), and JD degrees, all from Harvard. He was awarded honorary doctorates from the University of Munich (2006) and the University of Marseilles (2008).

Olivier Nana Nzepa is the Project Coordinator for Africa Management. Dr. Nzepa is an ICT lecturer with the Advanced School of Mass Communication (University of Yaounde) and Professor/Consultant in Public Management at the Institut Superieur de Management Public (Yaounde—Cameroon). His involvement in the World Summit on the Information Society (WSIS) includes the following positions: General coordinator of the African Civil Society for Information Society (ACSIS); Africa Region focal point on the International Civil Society Bureau for the WSIS; Member of the Africa Bureau representing civil society; Member of the RALO, Africa; Non-Commercial Users Constituency (NCUC) and At-Large Committee of ICANN; and Editor of a French quarterly dedicated to the Information Society titled "Le Defi Numerique." Dr. Olivier Nana Nzepa holds a Ph.D. in Communication from University of Montreal (Canada); an M.A. in International Relations from ENAP (Montreal); an M.A. in Business Management (H.E.C. Montreal); an M.Sc. in Communication (University of Montreal); and a Bachelor in Journalism from University of Yaounde (Cameroon). Dr. Nzepa is based in Cameroon.

Judith O'Neill is Chief Operating Officer and General Counsel of CMAS Holdings LLC, a cellular message alert system innovator. She has practiced law in the United States and internationally for 37 years, spending much of her time in the developing world, with a concentration on Africa, the Caribbean, and Latin America, spending the past two decades focusing on issues of privatization, regulation, and investment into the telecom markets in sub-Saharan Africa. She has worked as well in Asia, Eastern Europe, Oceana, and the Middle East. She concentrates her practice in telecommunications and new media/ICT sector, working on all aspects of regulations, sector structure and competition, joint ventures, privatizations, policy, transactions, and contracts. Ms. O'Neill has acted for carriers of all sizes, governments, suppliers, financing sources, and accounting/consultancy firms in telecommunications and ICT capacity building and development as well as on new media matters. Ms. O'Neill lectures all over the world on the legal, regulatory, strategic, and business aspects of the telecommunications/new media and ICT sectors. Most recently, she has been involved in the growth of next-generation broadband, SMS patented technology on location-based emergency alert systems; VoIP deployment and its role in the converged ICT and entertainment sectors globally, as well as with changing regulatory structures as they will impact market entry and business strategy predominantly in the United States and Africa; and development of ICT laws and policy in emerging economies, particularly in the Caribbean and Africa. Ms. O'Neill has worked in sub-Saharan Africa for more than 20 years and has worked and lectured in more than 80 countries globally.

Martin Phelps, Director Strategy and Market Development, EMEA-AT&T Business, joined AT&T in 1995, and his career in IT and communications spans over 20 years. Since then he has held various positions in Sales, Marketing, and Strategy, initially managing the client relationship issues for some of AT&T's largest customers. He was previously Global Product Manager for voice services with responsibility for all market-sensing and service launch activities for IP convergence, outside of the United States. In his current position Martin is responsible for the development and implementation of the AT&T strategy in Europe, the Middle East, and Africa (EMEA).

In this role Martin looks for new ventures and partnerships, new propositions, and new markets in which to support the multinational corporations that comprise AT&T's existing and prospective customer base.

Prior to AT&T, Martin held senior sales positions with IBM and BT, also managing large, complex clients. He began his career in the oil industry with Mobil Oil Company Ltd in London.

Martin holds a B.Sc. in Engineering from the University of Southampton in the UK as well as an M.B.A. from Henley Management College.

Mr. Brahima Sanou has 30 years of experience in the telecommunication and ICT sector. He has been working for 12 years with ITU, 10 of which he served as Head of the ITU Regional Office for Africa, a region with the largest number of countries and facing the greatest and diverse challenges in terms of development (28 out of the 48 LDCs in the world) in a multilingual environment (English, French, Spanish, and Portuguese). He therefore has a good understanding of the unique development needs of each country and region beyond common challenges, which will allow him to make sure that BDT is responsive to the common and individual challenges across regions and countries. A trained telecommunication engineer, he also holds a postgraduate diploma from the Center for Financial, Economic and Banking

Studies of Paris, which gives him the proficiency to handle economic and financial aspects of telecommunication/ICT development issues. He brings with him vision and a lot of energy to make BDT a truly operational arm of ITU, delivering ICT solutions for sustainable socioeconomic progress.

Christoph Stork, Senior Researcher, University of the Witwatersrand, Project: ResearchICTafrica.net.

Responsibilities included managing research programs, conducting quantitative and qualitative research, and lecturing and supervision of Master's and Ph.D. students.

Senior Researcher, Namibian Economic Policy Research Unit.

Responsibilities at NEPRU included managing research projects and supervising research teams, publishing in peer-reviewed journals, organizing and facilitating workshops and conferences, fund raising, and representing NEPRU in the media (TV, radio, newspapers). Research fields included financial economics, private sector development and trade, economics of ICT, and good governance and democracy.

Researcher at heart with the drive to know more everyday specialties: Telecommunication Policy and Regulation, Mobile Banking/Finance.

Glossary

3D Three dimensional
ADSL Asynchronous digital subscriber line
B2B Business-to-business
B2C Business-to-consumer
CAPEX Capital expenditure
CDN Content delivery network
CSP Communication services provider
EDGE Enhanced data rate GSM evolution
GPRS General packet radio system
GSM Global system for mobile communications
HD High-definition
HSPA High-speed packet access
IP Internet protocol
LTE Long-term evolution
NGN Next-generation networks
OPEX Operational expenditure
OTT Over the top
P2P Peer to peer
QoE Quality of experience
QoS Quality of service
ROI Return on investment
S&PEs Service and performance enhancers
SP Service provider
TCO Total cost of ownership
TIC Transparent Internet caching
W-CDMA Wideband code division multiple access

J. O'Neill et al. (eds.), *Broadband as a Video Platform*, The Economics of Information, Communication, and Entertainment, DOI 10.1007/978-3-319-03617-5,

The manufacturer's authorised representative in the EU is Springer Nature Customer Service Centre GmbH, Europaplatz 3, 69115 Heidelberg, Germany. If you have any concerns regarding our products, please contact ProductSafety@springernature.com

Printed and bound by CPI Group (UK) Ltd, Croydon, CR0 4YY
25/09/2026
02222841-0015